DAVID CHRISTOPHER SHOCK

marketing architectural and engineering services

marketing architectural and engineering services

SECOND EDITION

WELD COXE

VNR VAN NOSTRAND REINHOLD COMPANY

Manufactured in the United States of America

Published by Van Nostrand Reinhold Company Inc.
135 West 50th Street
New York, New York 10020

Van Nostrand Reinhold Company Limited
Molly Millars Lane
Wokingham, Berkshire RG11 2PY, England

Van Nostrand Reinhold
480 Latrobe Street
Melbourne, Victoria 3000, Australia

Macmillan of Canada
Division of Gage Publishing Limited
164 Commander Boulevard
Agincourt, Ontario M1S 3C7, Canada

15 14 13 12 11 10 9 8 7 6 5 4 3

Library of Congress Cataloging in Publication Data
Coxe Weld.
  Marketing architectural and engineering services.

  Includes index.
  1. Architectural services marketing.   2. Engineering
services marketing.   I. Title.
NA1996.C6   1982   720'.68'8   82-8340
ISBN 0-442-22011-1   AACR2

PREFACE

The preface to the first edition of this book, published in 1971, stated its intention "to lift the veil of mystery which has too long surrounded the promotion and new business development practices of architects, engineers, and similar professionals."

Apparently it succeeded. At this writing a decade later the ethical restraints which had masked the issue for generations have generally been swept away; the leading professional societies have begun sponsoring widespread seminars to teach marketing techniques to their members; there is a growing literature (see bibliography) that is giving the marketing of professional services all the status of a discipline of its own; and there is a large and growing organization, the Society for Marketing Professional Services, that is committed to making the craft of its members a profession of its own.

So why a new edition? The first edition pointed out that there are no significant differences in the way various successful architectural and engineering firms market their services. Getting commissions for such services is not a matter of proprietary tricks of the trade. Rather, it is the clients who seek architectural and engineering services who set the rules by which individual firms must compete for consideration.

In the decade since the first edition of this book a whole generation of clients has evolved who have had experience with the new, more competitive processes of choosing architects and engineers. Their response has confirmed a majority of the marketing techniques and strategies that were being practiced a decade ago, but the experience of such a large body of clients has also led to some changes and many additional approaches which are now applicable to the development of new business.

This edition endeavors to bring all this experience up to date. It has also been greatly expanded in order to address the management considerations that are now a part of the marketing effort in every firm that has a serious marketing program.

The material is organized in four sections:

The first section, Rules of the Game, defines the principles and the ethics, and is intended to orient professionals to how their needs for marketing fit in the context of business promotion and communication in general.

Section II defines The Business Development Process. It will serve its purpose of it helps the reader recognize the sequence of the selling process so that, in any given situation, professionals can recognize what in the process is going on, and what techniques are appropriate to respond effectively.

Section III, Tools of the Trade, is an explanation of the communications technology that is applicable to support of the professional's marketing efforts.

The final section, Making it Happen, entirely new to this edition, deals with the organization and management of marketing in professional firms.

A major addition to Sections II, III, and IV are actual Case Studies of successful marketing techniques applied by architectural and engineering firms across the country. These cases were originally published in *the Coxe Letter,* a newsletter of marketing techniques edited by the author between

1974 and 1980. The experience of these real firms, cited in the Case Studies, serves to validate the techniques and principles cited in the text. The great majority of the cases selected for reprinting involve smaller firms, and are cited particularly to demonstrate that successful marketing is by no means the sole province of large design organizations. Although the lessons of the cases are timeless, astute readers will notice that many of the people and organizations cited have evolved over the years. For historical accuracy, the actual dates of publication of the case studies are given in the index.

Readers will also notice that a preponderance of the examples used to illustrate the text, as well as the Case Studies, are taken from architectural rather than engineering firms. This is not due to any bias or lack of exposure by the author. It is historical fact that architectural firms experienced the new marketing climate in professional practice almost a generation earlier than did engineering firms. This happened chiefly because the majority of architects' clients are lay persons while most engineers' clients are other engineers. The latter historically had spread the work around among favored firms rather than face competitive choices. It took an extra decade before these engineer clients began to embrace the competitive marketing practices of the lay sector. By the time of this writing, however, the engineering marketplace has become just as marketing oriented as the architectural marketplace, and engineers are rushing to adopt and learn the same practices that architects embraced 10 to 15 years earlier.

This volume does not attempt to be a comprehensive how-to-do-it for all time, for selling is always a dynamic, evolving process, and the methods of selling must be flexible to keep pace. Neither can it be wholly applicable to the marketing needs of every professional firm, for the concepts of practice always differ from one professional to another, and each firm must find the marketing strategy that suits its style.

This material will serve its purpose if it encourages professionals to recognize the importance of marketing as an essential activity in the pursuit of their goals, and helps them understand how to apply the best marketing principles and techniques to their practices.

WELD COXE

Philadelphia

ACKNOWLEDG-MENTS

This volume acknowledges its lineage to those cited in the first edition, plus: David Vachon, Carl McConochie, and Thomas Harack, who, as staff writers for *the Coxe Letter,* researched and prepared many of the Case Studies reprinted herein; Michael R. Hough, publisher of *A / E Marketing Journal,* successor to *the Coxe Letter,* for permission to reprint those cases; Cindy Feldner, for so ably assembling the final manuscript; the many clients of The Coxe Group, management consultants, who have allowed the author to learn so much at their expense; and most of all, this edition acknowledges the members of the Society for Marketing Professional Services who, by their day to day actions in the marketplace over the past decade, have created the body of experience which validated the first edition.

CONTENTS

marketing architectural and engineering services

Rules of the Game

1

CAN THE PROFESSIONAL SELL?

"It's rather simple: The whole purpose is to get the job! It's the how that creates the problem."
Marketer Charles E. Edmonds

The day has passed when the nation's need for professional architectural, engineering, and planning services could be met by many small firms each serving a handful of clients obtained through personal friendship and contact. In the first place, the number of clients has multiplied tremendously, and more and more they are corporate or committee in nature. This means that the selection of professionals becomes a group decision based on analytical processes, rather than individual patronage. Secondly, the client is more aware of the benefits of quality rather than mere quantity, and this is creating a huge market for the services of the most able professionals. Few

3

of these outstanding practitioners—or their clients—are any longer born to "the establishment" where inheritance determines who gets the opportunity.

Thus it is that professionals have more incentive than ever before to embrace the process of new business development.

Although the demand for quality professional service has never been greater, it is more difficult than ever for the professional to get the attention of the prospective client long enough to present a sales message.

There is a growing awareness that perhaps the most difficult challenge of this age is how one person can capture the attention of another long enough to expose them to an idea.

Selling professional service is a matter of communication, and communication—the process by which ideas are exchanged—is a decidedly 20th Century word. It is said that one fifth of all knowledge has been discovered in the past decade. All man's knowledge accumulated prior to 1950 is projected to amount to but 8% of what will be known by the end of this century.

To cope with the consequences of all this, modern society has developed a whole range of new techniques. Marketing, advertising, promotion, public relations, merchandising, propaganda, and publicity are the terms and the tools.

The process is communication.

The objective is to project a fresh idea and have it acted upon. This is the essential goal of all communication today, whether the message is on behalf of a manufacturer, retailer, politician, educator, engineer, or architect.

Thus, the communication environment in which architect, engineer or other design professional must compete is little different from that in which automobiles are sold, groceries are bought, or elections won and lost.

It matters not that the professionals are selling a service rather than a product, that they may or may not be ethically free to use every technique in the book, or that they may need only a half-dozen commissions a year while Ford must sell a million cars.

In a very real sense, the process by which an architect or engineer reaches a prospective client and obtains a commission is little different from the process by which Ford or BMW sells the professional an automobile. In both cases, there are four basic levels of communication involved.

Recognition of Need. If a person who needs transportation thinks a bicycle will do the job, he or she isn't going to shop for a car. It is therefore no accident that the automobile industry makes sure that America's teenagers are clamoring to own cars even before they are legally old enough to have a license.

The situation is just the same in professional services. If the client who wants a custom house first calls in a builder, it is obvious that architects have incomplete communication. The same is true when legislators introduce bills for in-house design of all engineering work. The individual architect or engineer can do little to overcome this kind of uninformed audience. This is the job of the societies such as AIA, ACEC, NSPE, ASLA, AIP, etc.

Class Identification. When a customer decides to shop for a low-priced, gas-saving automobile, will they think of 2, 3 or 4 domestic manufacturers, or 5 or 6 imports? On that question hangs the fate of companies such as Chrysler, American Motors, Renault, Mazda, and several others. A great deal of these companies' promotion is aimed at convincing prospective car buyers to at least consider their brands in addition to the "major" manufacturers.

Individual architectural or engineering firms face very much the same problem. When the local school board or highway department sends out invitations for professional service proposals, who will be on the list? Professionals must be sure their firm is counted in when all the firms they consider equal are invited to submit proposals.

Competitive Preference. After the automobile buyer has shopped all three, or four, or eight, there will be preferences. One car uses less gas. One rides more smoothly. One looks better. One is cheaper. How does the buyer know? Communication. Which will be bought? That won't be decided without salesmanship.

The professional is selected in much the same way. Initially a firm may be one of a dozen or more under consideration. Then the choice will be narrowed to two or three. A critical challenge to the professional's communication program at these stages is how an individual firm can have an identity different from and superior to its competitors.

Selection. "Chevrolet probably has the better car, but the Ford dealer is closer to my office and service will be easier to get." If the Ford dealer intended the buyer to reach that conclusion, good salesmanship was involved. If the Chevrolet dealer offers free pick-up and delivery for service, but did not get this message across, there was poor communication.

How one professional firm wins a commission over its closest competitors is an equally complex—and sometimes capricious—process. It involves information, personality, strategy, and some luck—all of which are elements of communication. The name of the game is salesmanship.

Experienced salespeople recognize these steps in the sales *process* instinctively. It is the difference between the *process* of communication and the *techniques* that is significant for professionals. It is generally accepted today that the projection of each idea, product, or service requires an individual communications system. The techniques used in a particular system will vary considerably according to the nature of the idea and the audience to be reached. But the techniques, or tools, are simply the means. The end goal is always communication.

The professional who is successful at selling services is first of all a skilled communicator. Salesmanship takes skill and work. The heads of most successful professional proprietorships will acknowledge that they spend a high proportion of their time "getting work." Historically, many professional partnerships have been formed on the basis that one partner will "bring in the work" while another "designs it" and possibly a third "produces it." These associations are founded in recognition of the fact that all professionals are not endowed with equal aptitude for selling, and so the one with the natural inclination for salesmanship tends to specialize.

What is often not recognized, however, is that natural endowment is by no means the only way to become a successful salesperson. Professionals can predetermine their marketing goals and then can learn to reach out and land work. In simplest terms, that is what this book is all about.

There is no mystery to this process and, despite some genuine opinion to the contrary, there are no significant trade secrets. On the increasing occasions when professionals compare notes about marketing practices, the one fact that stands out most clearly is the marked similarity among all successful selling approaches.

It is also important to recognize who should be responsible for marketing in a professional firm.

Many of the larger professional firms in the country today employ a variety of nonprofessional people in their business development programs. Some use outside lead finders, or "bird dogs." Others have marketing directors/managers, and public relations specialists on staff. Some more have PR or advertising agencies available on retainer.

All these people can and do have a useful place in the successful business development program, as will be discussed. But the author does not look for the day when the entire selling process in any professional firm will be handled by laypersons—however qualified they may be in their own fields.

The finite difference between professionalism and commercialism is the essential one-to-one relationship between client and professional. Design itself is an art of communication, and no sales process can—or should—tamper with this. Sales specialists and sales technology can be used to expedite achievement of the professional relationship, but never to replace it. The ultimate salesperson in every professional firm will always be the principal in charge.

The author is convinced that improved salesmanship among professionals has done far more to enlarge the total market for these services than it has hindered any individual firm by benefiting its competition.

This volume has been prepared for professionals who recognize that marketing professional services is their responsibility, that the process is no mystery, and who are ready to accept the challenge to master it.

Can professionals sell their services?

They've been doing it as long as there have been professionals.

2

IS IT ETHICAL?

"The refraining from or the employment of advertising is the clearly defined difference between a reputable physician and a quack."
American Medical Association ethics, 1910

"Are there any among us unwilling to acknowledge that the advertising question is but one of degree, and that the practice of architecture owes its very existence to publicity? In the interest of progress may it not be high time to recognize openly that the place for one's light is on a candlestick rather than under a bushel?"
Thomas R. Kimball, FAIA, 1918

"It's basic. Our mission is to design and build to serve people. If we do not have projects, we cannot do architecture."
Paul Kennon AIA, 1976

As the foregoing quotations illustrate, the issue of where marketing fits in professional practice has been debated for a long time. Until the early 1970s the different professional societies tried to answer this question for their members by writing elaborate codes of ethics which spelled out in sometimes explicit detail how design professionals could (or could not) go about offering their services to a prospective client. The history of all this was documented in the first edition of this book, published in 1971, and those who are curious about the evolution of the ethics and the extent to which different professions tried to prescribe the rules of marketing may find it interesting to read.

The relevance of the ethical question changed radically in the mid-1970s, however, when the U.S. Justice Department went to court to claim that the marketing canons of all the so-called learned professions were illegal under the antitrust laws because they limited the ability of professionals to compete for work.

Some of the design professional societies were quicker than others to accept the principle and rewrite their codes. One, NSPE, fought the change at great expense all the way to the U.S. Supreme Court before it conceded. By the end of the 1970s, however, it was universally accepted that, like it or not, professional societies do not have the legal right to regulate how their members market their services. The U.S. Sherman Antitrust law is very clear: "Any act" that tends to restrain trade is illegal. *Any act.* In the end, the professional societies had to concede that it was as simple as that.

Once conceded, however, there then began within the professions considerable debate about what of their members' conduct they could or should try to regulate. Some societies tried to continue to promulgate mandatory (and, therefore, enforceable) ethical standards by remaining completely silent on the marketing areas. Other professions tried drafting "voluntary" statements of principles which included some advice about what to do and not to do in promoting one's services. At this writing several of the major professions were still swinging back and forth between the voluntary vs. mandatory approach, but it appears in time the issue will become moot.

To understand why this is so one must accept that there are really two very different questions involved:

1. Is marketing ethical?

2. Is marketing professional?

In the ethical sense, it is now abundantly clear that it is not legal (or, therefore, "ethical") to limit the freedom of professionals to solicit work. Accepting this really answers the whole question, because to date no one has come up with marketing regulations that are not anticompetitive.

The *professional* question, however, is an entirely different matter. The fact is that architects, engineers, and other design professionals don't set the rules by which they compete: Their clients do. If clients for design services wanted to be sold by commission agents, you can bet there would be all sorts of commission salespeople on the payrolls of design firms. If clients took seriously every proposal that they replace their present engineer or architect with a new firm, there would be lots of supplanting going on. If clients were turned off by receiving unsolicited brochures, newsletters, and advertising from design firms, it would quickly stop.

The point is that most buyers of design services have their own very clear view of professionalism. If what architects or engineers do to market their services falls within the client's definition of "professional," they will deal with those firms. If, on the other hand, what an architect or engineer does to sell themselves appears to the client as manipulative, slick, evasive, or otherwise similar to the classic image of a used car lot, 99 out of 100 clients won't buy it.

The clients' reception, then, tends to define those marketing techniques that are professional and those that are not. It is interesting to note that since the government forced suspension of the "ethical" rules about marketing design services, one or another firm has experimented with all sorts of previously proscribed marketing techniques—commission agents, self-laudatory promotion, franchise schemes, bait-and-switch fees, etc., etc. In almost every case the experiments have been dropped. Clients just don't buy professional services on used car lots.

What really counts in the marketing of professional services is whether a given technique or approach reinforces the professionalism of the seller to the client. In practical fact, most of the techniques prohibited by professional society ethics prior to 1970 are not very useful because they don't sell in the marketplace. Those few that do help haven't lowered the standards of professionalism one bit.

Is marketing ethical? If it is *professional,* you bet it is ethical.

3

MARKETING VS. SALES—DEFINING TERMS

"Marketing has to do with the creation of credentials; sales is going up and knocking on doors."
Marketer William E. Flaig

Marketing deals with what you sell. *Sales* deals with how you sell it.

The terms, *marketing* and *selling,* are often used interchangeably in the professional world—and in this book. In the literal sense, however, there is a distinct difference between the two as defined above. It is important to understand why this difference is less applicable to professional services.

At the fundamental level, the distinction begins with the difference between a product and a service. A product, in the sense of, for example, an automobile, is a predetermined object which one buys or doesn't buy according to how the features it offers apply to one's needs. One may be

11

offered numerous options to make it more or less applicable, but once Ford's marketing department has established the criteria for this year's model, and the assembly line is cranked up to make it, the customer must take it—or leave it—generally in that form. In the product world, marketing determines what is to be sold, and selling is a clearly distinct function which tries to persuade as many people as possible that the established model is just what they want.

A service, on the other hand, is frequently marketed and sold at the same time. The cusomter who wants a bridge or a building designed is originally more concerned with the process by which the engineer or architect will design it than with the end result. Considerations such as the firm's professional response to the program of the project; its approach to project management, and ability to meet budgets and schedules; and the personal chemistry between buyer and seller are the elements being purchased. As the skillful design firm adapts what it can do to the particular needs of the assignment at hand, and puts together a persuasive proposal demonstrating why its approach is just right for this particular project, it is marketing and selling simultaneously.

In the special climate in which professional services are offered and commissioned, it is the author's observation that the factors which lead to the most success in marketing divide approximately as follows:

70% to 80% of the success of design firms is due to what they sell;

20% to 30% of the success, at most, is due to how they sell it.

In this context it has been demonstrated time and again that firms which offer a truly marketable service will be found by clients and will have lots of good assignments whether they are doing any active selling or not. Conversely, it has been equally demonstrated that firms which learn to sell but don't have much to offer may win a job or two, but, unless they deliver viable services, they are quickly found out by the marketplace and become less and less saleable.

Thus, although this book deals heavily with the "how" of selling professional services, it is interchangeably a marketing book because it assumes professionals understand what selling cannot do for them. Namely, it can't begin to help get work for a firm that is not offering marketable services. If you are doing all the selling right and are still losing to competitors, try working on the "what" you are selling.

At the same time, it must be emphasized why learning to sell is essential for that great majority of firms which do offer highly marketable services. Today's clients have become very sophisticated at the process of sorting out and identifying several highly qualified firms to consider before selecting one to provide the desired service. And despite the ego of many designers who would like to believe that they have no equals, the short lists most clients prepare include several firms who could do an equally competent job with the assignment.

Who will get the job?

More often than not, it goes to the firm that best knows the definitions of marketing *and* selling.

4

WHAT IMAGE WILL YOU SELL?

"When service is superior, public relations is superior. If poor service is being rendered no amount of propaganda will produce good public relations or convince dissatisfied customers their service is satisfactory. All the public relations publicity in the world will not make bad public relations good."

Newspaper Editor C. A. Turner

Image is a thoroughly overworked word in the communication world these days. It has been abused both by communicators and businessmen, and the result is a good deal of confusion about who is responsible for the company image and what should be done about it.

The most unfortunate misconception is that expressed in titles like *The Image Makers*—the myth that "Madison Avenue" can change a "bad" image into a "good" image. Nothing is further from the truth. Public opinion is a synonym for the word "image" and, in the long run, the truth about a product or business will be reflected in the opinions held by its principals, employees, and customers.

Every business has an image from the moment it opens its doors. The founders of a given business can, by their approach, endeavor to shape their initial image. A new professional firm may, for example, wish to be known for "design," or for "highways," or for "comprehensive service," or any one of numerous special qualifications or talents. It is up to the principals to decide what image they want, and in this context, there is no such thing as an image being "good" or "bad."

What really matters is whether the image is "accurate" or "inaccurate." A professional firm which wants to be counted in among the so-called "design firms," but only gets invited to compete with those it considers of inferior talent, is having trouble with its image. But the first question to be asked is not: How do we change it? Rather, the starting point should be: How did we get the image we have?

In practice, every professional will be concerned with (and affected by) two different images: First—and most important—is the image of the profession as a whole. Second is the image of his or her own practice.

IMAGE OF THE PROFESSION

Individual professionals cannot escape from the image of their profession, whether they like it or not. If people believe that engineers are stodgy and architects are expensive, no single professional (no matter how creative or cost-conscious) is going to change that on their own.

If a professional does not agree with the image of the profession, the place to begin efforts to change it is through their professional society. There are many dramatic examples in which communication and public relations techniques have been successfully applied to shift public opinion where the cause was sound and just.

The basic tool for all this is public opinion research—the closest thing we have to a scientific method for measuring an image. The results of public opinion research are invaluable to planning communication programs where the goal is a shift in public opinion. And at the end of a communications campaign, opinion research can measure what has been accomplished.

A good example involves the railroad industry in the state of New Jersey. Some years ago property taxes on railroads within New Jersey were the highest in the nation. The railroads were in financial trouble and wanted to seek tax relief, which required an act of the legislature.

As a first step, the railroads conducted a state-wide opinion study. It revealed that less than 3% of the people of New Jersey believed railroads were overtaxed. Among opinion leaders—those in business, community affairs, or government who, by their station, tend to influence the opinions of others—the proportion was only 8%.

On the basis of this study the railroads aimed an information campaign at opinion leaders, using speeches, newspaper publicity, and direct mail.

Railroad executives actively sought opportunities to speak before service clubs and business groups throughout the state—and made more than 150 speeches in a year. The subject: Railroad taxes. Publicity people saw that each speech was reported in local newspapers. Favorable articles and editorials were then reprinted and distributed by direct mail to lists of opinion leaders and legislators.

A year after the campaign began a second public opinion sampling was made. The proportion of general public who now believed railroad taxes were too high was up from 3% to 8%. Among opinion leaders the proportion had jumped from 8% to 30%. With this background of public understanding the railroads sought introduction of the desired legislation, and it was passed with bipartisan support.

This rather selective example is used to demonstrate that effective image campaigns need not be limited to consumer topics and mass audiences. It is perhaps particularly significant that while the railroads were successfully seeking tax relief in New Jersey, their general image, because of degenerating passenger service, was anything but "good." And the tax story did not attempt to change that image—nor could it have—because the service was inferior.

The writer believes that the design professions could benefit greatly by application of opinion research analyses to the planning of their communication programs. Both architects and consulting engineers in particular have been complaining for a generation that: "The public does not understand us." But despite years of talk at national professional levels about "getting public relations," little has been done to improve the image of the professions.

One of the basic problems is that too little is known about what those who may buy professional services actually think about the design professions.

For example, to demonstrate the probable disparity between what architects *believe* their clients are thinking, and what the clients really think, questionnaires were sent several years ago to members of the Bucks County, Pennsylvania, AIA Chapter and to 200 members of a Parent-Teacher Association in a white-collar Philadelphia suburb.

One question asked the public to assume they were to launch a building project for themselves—either residential, commercial, or institutional. Eight individuals in the building process were listed in the question:

Architect

Banker

Engineer

General contractor

Lawyer

Materials supplier

Realtor

Subcontractor

The public was asked to tell in what sequence they would expect to deal with these eight in order to get their building built.

The purpose of this question was to find out how well the public understands the architect's role as the owner's principal representative and counsel throughout the entire building process.

The architects were sent the same questionnaire but were asked to guess what they thought the public's answers might be.

The architects correctly guessed that the realtor would take first place. But this was not the important answer so far as the architect's image is concerned. Most of the surveyed architects—39 percent—guessed they would be the fifth person on the list to be contacted. Another 22 percent guessed that the architect would be fourth, and 17 percent were optimistic enough to think the architect would be the third contacted.

The public disagreed. A large 39 percent of the public said they would call the architect second, and 21 percent ranked them third. Only 4 percent of the public ranked the architect as low as fifth—where most of the architects thought they would come into the picture.

In a subsequent question, the public was asked to name the one single individual in whom they would want to have the greatest confidence to assure the success of the project. The same eight choices were given. The question is really more important than the previous one because if the architects do not rate the confidence of the public they can never hope to survive as an independent profession. In answering this question the overwhelming majority of the architects—55 percent—guessed that the public would want to have greatest confidence in the general contractor.

On the contrary, an overwhelming majority of the public said it would want to have greatest confidence in the architect. Only 28 percent of the public gave highest ranking to the general contractor.

It must be made clear that this survey was not conducted to prove anything valid about the architect's image. The survey simply demonstrated that a small group of people in suburban Philadelphia have an opinion about architects, and that AIA members in a nearby Chapter thought these people would have a quite different opinion. It would be no more valid to base a public relations program on this survey than to lay out foundations for a building after a single test boring on a 100-acre site.

In 1967 the AIA commissioned a serious opinion research study entitled "A Study of the Architect's Image among Clients." One question in the study reveals how clients evaluate the services of the architect. The question asked,

> "Which of the following services that architects provide do you consider to be the most important?"

The answers, in rank order, were:

1. Providing functional planning.
2. Seeing that the contractor adheres to specifications.
3. Creating aesthetic design qualities.
4. Developing building specifications.
5. Acting as the owner's agent.
6. Maintaining control of the budget.

These answers suggest so many other questions that they could well be the basis for an entire new opinion research study. At the very least, however, it is apparent that the *satisfied client* hires the architect for reasons considerably different from the ideals taught in architectural school.

A question and answers such as the above may lead to action in either of two quite opposite directions. On the one hand, the architects may decide to try to change the opinion of their clients. On the other hand, the architects may decide to change themselves in order to bring their services more in line with what the clients seem to want.

If professionals stop guessing and find out what their public image really is they will have accurate information on which to base programs for their professions. Individual practitioners may also learn about the services they should or should not be stressing as they sell their services. The point is that research techniques are available which can guide professionals in responding to the image of their professions.

IMAGE OF THE FIRM

The public image of engineers or architects as a group creates the market for professional services. But it is the image of an individual firm that will capture a share of that market. The same principles of image analysis apply when an individual firm is planning its own marketing strategy. Foremost is the need to be sure that your view of yourself is shared by your clients and prospective clients.

When all is going well, and a firm has a good batting average in competition for projects plus a high ratio of referral commissions walking in, it can be assumed the firm's image is positive. This does not mean, however, that the principal's view of the firm's image is the same as the image held by clients. For example, an engineering firm that thought of itself as strong in land planning services because it had most of the work in town, found through an image study that it got the work because it was cheaper, not because it was thought of as particularly strong professionally. Since the firm was having trouble making money, the image study demonstrated that it had a much larger marketing problem than was originally recognized. The converse can also be valuable: An image study that confirms a firm is as strong as it thinks it is can be a healthy reinforcement for continuing existing marketing strategies.

More frequently, however, image studies are used by design firms to address specific concerns. Three specific cases are worth citing.

Case A. The founder of a small architectural firm in the Southwest recruited three young partners to take over his practice as he approached retirement. After two years of trying, however, the younger three, despite excellent qualifications, were having great difficulty winning work in competitive situations.

An image study was conducted among some three dozen of the firm's past clients, recent prospective clients where the firm had lost to competitors, and potential future clients. The feedback was surprising. While the founder was widely admired for his professional convictions and his role in encouraging civic improvements, he had developed an image as being opinionated, inflexible, and hard to work with. Clients were reluctant to give work to the firm so long as he was perceived to be in control.

The marketing solutions: Change the name of the firm to give all four principals equal billing; elect one of the younger principals as chief executive officer with appropriate wide publicity; and omit the founder from further interviews while presenting his role as an "emeritus" resource when appropriate and necessary. Once these actions were taken, the fortunes of the firm changed dramatically, and several major commissions were won. In due course the founder was successfully bought out and his name was retired from the firm's letterhead.

Case B. An established and rapidly growing engineering firm trading under its founder's names (e.g., Able, Baker, Charlie, Inc.) considered changing its name in order to deemphasize the role of the founders and help spread the marketing identity among a larger group of project leaders. The firm decided to call itself by its initials (e.g., ABC Engineers, Inc.) and began planning a wholesale identity change—new stationery, forms, brochures, job signs, etc.

Before the change was implemented, however, an image study was undertaken to examine several marketing issues, and the name change was included in the survey. There was immediate negative feedback. The firm's admirers said that initials (ABC Engineers) had no personality and would not be seen as a credible identity—anyone can call themselves by initials. Conversely, the study revealed very strong good will for the original name of the firm and an overriding belief that using people's names is "more professional."

Needless to say, the name change program was reversed and the firm adopted other marketing communication strategies (a newsletter, professional public appearances, authoring technical papers, publicity, etc.) to give identity to its new group of marketers.

Case C. A large architectural firm in a major growing city found itself barely holding its own in size as its six major competitors took off with the community's growth.

The firm commissioned an image study to compare its position with its competitors. A selected sample of public and private clients, contractors, lenders and construction consultants was contacted and asked to rank the seven leading firms in town (including the sponsor of the study) by their perceived strengths. One such question asked:

Can you rank the following seven firms by their strength in construction administration services?

The sponsor of the study found itself dead last on this question.

To address this perceived shortcoming, the firm decided to create a separate Construction Services Department, headed by an experienced construction field representative. This move was duly announced to the firm's clients, published in its external newsletter, emphasized in proposals, and, of course, experienced directly by construction companies working on the firm's projects.

Two years later the image study was repeated and the same questions asked. This time, on the question of the strength of its construction services, the sponsor had risen to a three-way tie for first place among its six competitors. Equally important, the second study showed the firm's image in several other problem areas (e.g., high-rise office experience, and design innovation) was still weak even though the firm had tried to talk its way out of these deficiencies with PR and special brochures during the ensuing two years. The firm in fact had taken no action in these weaknesses—it just talked about its own self image and this hadn't worked. As a result of the second image study the firm adopted a plan to recruit a recognized designer of high-rise office buildings into the firm.

HOW IMAGE STUDIES ARE DONE

Image studies such as those cited above can be conducted by marketing and market research consultants, or, in many cases, a design firm can conduct its own study.

A typical image study involves two types of interviews:

"Open" interviews where the sponsor of the study is identified and the objective is to get a candid appraisal from people who know the firm well—usually clients and close professional or trade contacts.

"Blind" interviews, where the sponsor is not identified, and sources are contacted to obtain unbiased and objective answers. Blind interviews are most frequently used when checking comparative rankings such as are cited in Case C above.

A comprehensive image study will generally involve both types of interviews: some open interviews to get direct and candid information without chance of distortion through misunderstanding, and some blind interviews to check up on competitive images and to verify the extent of an image beyond those directly familiar with a study sponsor. Although samples as large as 50 or 100 interviews are sometimes suggested, a valid image study for most firms can be accomplished with as few as 6 to 12 open interviews and/or 15 to 20 blind interviews.

Some firms tend to be anxious about open interviews, either because they are not convinced the people interviewed will be honest, or because they fear the answers will be embarrassing and may jeopardize the relationship. Experience is just the opposite: People like to be asked to help. Consider how you would react if you received a letter like this:

Dear ＿＿＿＿＿＿＿

As part of an internal planning effort seeking ways to improve our services, we have asked a consultant to talk with several of our valued clients to get their perspective on how we are doing and what we might do better.

You will be hearing from ＿＿＿＿＿＿＿ of the consulting firm who will ask for an appointment to talk with you. We will be very grateful for whatever comments you can give about us. Please be candid. Your comments will be completely confidential.

Sincerely,

In practice, almost everyone likes to be asked for advice, and 19 out of 20 people who receive this type of letter from an architect or engineer are very willing to be helpful. Even more reassuring is the frequent direct feedback from interviewees complimenting the sponsors of such studies for being farsighted and asking about the experience because they would like to consider doing the same for their own organizations.

This virtually universal reaction to open interviews is also the key to doing an image study yourself. All it takes is a commitment—and the courage—to invite a client to lunch or meet and say:

> "We are interested in how we are perceived by those who know us best. I'd like to ask you to comment on some questions about us and our services, and I will welcome your candid feedback."

Abundant experience demonstrates that people given this opportunity to be helpful will do so constructively. There is rarely a problem getting them to be honest. The real issue is whether the architect or engineer asking the questions can handle the answers. Our paranoia makes such encounters risky for all of us. If we get defensive we blow the chance to make the client believe we really wanted the help. If defensiveness is going to be a problem, let a consultant do the study for you. If you are confident you can handle the answers, doing it yourself will inevitably strengthen your relationship with your client.

Whether you do it yourself or use a consultant, the key to a useful image study is knowing what questions to ask. Every image study needs to be tailored to the firm that sponsors it, and there are many different approaches that can be used depending on the particular underlying issues that the survey is intended to address. For those interested in constructing their own lists of image questions, the following may suggest useful approaches.

Open Interview Questions

In open interviews there must first be a set of questions to establish context. These would deal with the subject's experience in working with professional firms, the names of other firms with whom there has been direct experience, etc. With this background understood, the following can be helpful open questions:

How did you first learn about (the sponsor firm)?

(If a client.) What led you to choose them to work with you?

Have they lived up to your expectations?

What do you consider their strongest qualities and services?

What about the firm or its services might you like to see strengthened?

Has the firm changed in any ways during the time you have known it?

Are there services the firm does not offer which you would like to see added to their repertoire?

Would you recommend the firm to others? Are there any types of work for which you would not recommend them?

What firms do you consider competitors of (the sponsor)?

(For each firm named in the previous question.) What are their particular strengths? What are their shortcomings?

Of these competitors, which do you consider most similar to (sponsor)?

Of the competitors you previously named, which firms are most different from (sponsor)? What are these differences?

Is there any general advice you would like to give to (sponsor) which you think might help them better serve yourself and others?

Etc.

It should be clear than an image study composed of answers to these questions can lead to some valuable insights. It is always possible that one or another individual interviewee may have a particular perspective (or ax to grind) that will provide answers of limited value. But when a sample of six or more open interviewees are asked the same questions, and certain answers come through in a clear majority of the replies, that information can be considered a valid profile of the image of the firm that undertakes the study.

Blind Interview Questions

The principal value of blind interviews is what they can tell a firm about its image among people who know it less well—especially prospective clients with whom the firm would like to work and/or opinion leaders (such as other consultants or contractors) who may work with many similar design firms and can be expected to have useful comparative opinions.

Blind interviews must be conducted by market researchers who can represent themselves as working for an anonymous client and who will treat all individual answers confidentially. Typically such a researcher might identify their purpose by saying to the subject: "We have been asked to get references on five or six (architectural or engineering) firms that are active in your community (or field)." Then the questions might take these directions:

What firms would you suggest we consider?

(If the client is not mentioned.) Is a firm named _____ also active in this community (or field)?

Have you previously worked with any of these firms?

Can you tell me something about how large the different firms are?

What sort of work or project types are the different firms recognized for?

Have any of the firms undergone particular changes in recent years?

Can you say anything about individuals in the different firms? Do you know whom I might contact in each firm?

What qualities do you most look for when you select an (architectural or engineering) firm?

Which of the firms we have been talking about do you believe have those qualities?

If you were going to rank the firms you have mentioned for a (example) project, what order would you list the firms?

Is there anything I haven't asked about these firms that might help round out the references?

Clearly, blind image interviews can product useful information, but it must be interpreted with more caution than open interview data. Many contacts, when asked for references by a stranger, are reluctant to be negative. Frequently, these sources will simply be neutral or give "faint" praise. Therefore only the most positive statements are likely to be valid. Also, since blind interview data must be received without attribution to respect the confidentiality in which it is obtained, it is possible to overreact to those negatives which do come through. It is important to remember that nobody is perfect, and individual negative comments are not half as serious as a general pattern of criticism would be. Blind interviews have their greatest value in what they can tell you about your firm's position in relation to your competition. If you know your competition well enough—and every good marketer should know a lot about their competition—learning whose image is ahead and whose image is behind your own can be very useful.

PUTTING YOUR IMAGE TO WORK

The general point made earlier about image studies is the most important: It is the image of an individual firm that will have the most to do with that firm's share of its market.

Most professionals know the client does not entirely trust or understand them, but very few recognize this in the way they sell their services. They merely present themselves as a member in good standing of their profession, and suffer all the consequences of that image.

If a professional firm honestly believes it can provide better services than its competitors, it is very probable that it has a philosophy of practice considerably different from the others. Perhaps the organization is structured in a special way to foster coordination, creativity, or project continuity. Or the firm's estimators have techniques, qualifications, or training which make their cost projections more reliable. Whatever the special circumstances, the professional firm must actively articulate its philosophy if it is to enjoy a distinct image within its market. The most successful professionals, almost without exception, are those who are most capable of communicating an individual philosophy.

Even if one firm's philosophy is very similar to that of its competitors, the professionals in that firm should never be shy about articulating their beliefs. Competitors may not be saying it as well. And the prospective client, from natural ignorance, will choose the firm that makes its services more clear. The ultimate image of a firm, of course, is determined not only by how well it verbalizes its philosophy, but how well that philosophy works in practice. The firm that can prove it practices what it preaches will come out ahead of any firm that talks a lot but delivers little. How this can be applied in practice is discussed in detail in Section II.

What image will you sell? This chapter has tried to give a four-fold answer:

1. The image of one's profession is tied to each professional, like it or not;

2. The image of a firm or a profession can be tested, and communication techniques are available to shift an image if it is inaccurate;

3. The firm that stands out from the crowd sells an image of its own;

4. The validity of any individual image will ultimately be demonstrated in actual practice.

5

WHO IS THE AUDIENCE?

"The easiest way to get work is to get repeat work from satisfied clients."

Jerome M. Cooper AIA

A marketing program must always consist of three elements—a product or service, a method or medium of communication, and a customer or audience. As noted in the previous chapter, it is not the purpose of this volume to help define the product—each professional practice must decide for itself what it is trying to sell. When this is established, it is next most important to have a clear understanding of the audiences who will be aware of the professional's image and their relative roles in marketing.

The audiences that are of concern to the professional are far broader than the list of prospective clients. An architect or engineer in general practice has at least six basic audiences:

> Clients
> Employees
> The Building Industry
> The Profession
> Prospective Clients
> The General Public

The interests of each, and what the professional may want each to know, can vary significantly.

CLIENTS

By far the largest percentage of all commissions for professional services are awarded as repeat business from satisfied clients or through referrals they give to others. Thus, in the great majority of practices, the professional firm's image with its present clients will have more to do with its future business than its image with all other audiences.

It is the author's observation that very few practitioners have any valid idea how they are seen by their clients. For example, a firm in general practice, which constantly tries to maintain a diversified workload, will nevertheless tend to be thought of as specialists by clients. A hospital client of this firm will normally think of the firm as hospital designers. Thus, when the president of the hospital is asked to recommend professionals to the local school board the firm that is serving the hospital may be completely overlooked.

Client communication also plays a direct role in the execution of a professional commission. Many principals of medium size or larger firms complain that they cannot get their clients to deal with project managers. The clients, they say, insist on talking to the principal, even on minor matters. The answer to this is often another question: How well do clients really understand the professional's organization and the role of each person with whom they should deal?

There are three prime opportunities for client communication which should be a part of every professional practice.

Orientation and Briefing

Immediately after a commission is awarded there is an opportunity to gather the client's entire project staff and brief them on how the job will be served and with whom they will be dealing on the professional's staff. The normal process of programming, design, documentation, and construction is only vaguely understood by most lay clients. A good orientation at the outset can do much to build understanding and cooperation, particularly in sensitive areas, such as timetables. The client who is not told why the professional needs six months for schematic development will likely become impatient when there are no studies to see in a few weeks. But if the client understands that a detailed regulatory check is required, involving a bureaucracy that bottlenecks everything, the client will be more tolerant. Similarly, it is a good idea to show new clients an organization chart for the project, graphically illustrating where each person will fit.

Also, orientation is a good time to let the client know about all the services the firm is equipped to offer. Few firms are engaged at the outset of a project to render all the services of which they are capable. If no special effort is made to tell the client that the firm can perform, for example, space planning, interior design, landscaping, construction supervision, etc., the client may buy these services from someone else. One of the simplest places to sell ancillary services is through option clauses in the professional contract. A well written contract can be a very effective sales tool.

Project Presentation

Every client should have a clear understanding and appreciation of what the professional has done, and the ability to explain this to others. It is amazing how seldom this is so.

Successful professionals are usually quite adept at persuading clients to accept their solutions. However, the presentations of these ideas are often verbal, accompanied by graphic materials of a more or less technical nature. At the end of such presentations the client may be completely satisfied with the professional's work, but can be poorly equipped to explain the design to others. This is most unfortunate, for clients always have many audiences of their own and what they say to them about the professional's work can enlarge the professional's image.

Fund-raising brochures are a typical example. The client sends these to the most influential people in the community. If the fund-raising brochure carries a clear statement of what an architect has designed, it can influence a lot of public opinion. On the other hand, if the brochure carries only a reproduction of an "artist's sketch" and some floor plans interpreted by the client's public relations people, the architect will be lucky if the reader even remembers the firm's name (assuming it gets a "credit line" at all).

The opportunities are no less when an engineer designs a complicated interchange on the Interstate system, or when an architect designs a factory addition for General Motors. The client, in the normal course of business, will explain the design to literally hundreds and sometimes thousands of other people, many of whom are potentially important receivers on the professional's communication network.

If at the time of initial presentation, the professional provides the client with concise written and graphic summaries of the design concept, the client will use these as the basis for all later communication. If a rendering is provided to the client already framed, with the matte titled and containing the design firm's name, there is a 99% chance it will go upon the client's wall for all to see. If an elegantly bound brochure of model pictures is added, it will likely stay on the client's coffee table to be shown to visitors for months while the project is under construction. It is as simple as that. How this can be accomplished in practice is discussed in detail in Chapter 16.

Continuing Information

When the professional/client relationship achieves the qualities of true agency, lasting friendships are made. The client who is a friend is always interested in what the professional is doing, even years after the project is completed. Thus, professionals should not be shy about keeping clients informed of other projects they are designing or what kind of work they would like to do. Practical ways and means of doing this will be discussed in Chapter 19.

EMPLOYEES

All businesspeople today recognize that communication with their staff can have direct bearing on morale, productivity, and compensation. The term "employee communication," however, too often conjures an image of company house organs filled with statements by the president and social notes about the secretaries.

In professional offices, there are important additional areas of concern about employee communication, since professional staffs are regularly in contact with clients, public agencies, building materials salespeople, and other professionals. It is extremely important that the staff of a firm communicate to these audiences on the same frequency as would the principal.

Certainly no one questions this, yet it is common to find professional offices where two or more projects are being handled by different teams of people and neither has a very good understanding of what the other is designing, or why. Employees are an excellent audience before whom to dry-run project or interview presentations. Engineers in particular are notorious for neglecting this area. Setting up a system to assure that employees are briefed on the firm's work is really very simple, and can help give the staff a valuable sense of group participation.

Perhaps most important of all employee communication is the understanding each member of the staff has about the firm's professional philosophy—what it believes and how it approaches its assignments. Orientation at the time new employees are inducted is the best time to accomplish this, but it is surprising how few firms make a serious effort to adequately brief new staff members. A very effective way to do this is to set up in the firm's conference room a stock interview slide show with tape-recorded commentary, and play this for each new employee the day they are hired.

THE BUILDING INDUSTRY

The consultants, contractors, building material producers, and salespeople with whom the professional deals generally comprise one of the largest audiences. They are in and out of the firm's office every day—more often and in far greater numbers even than clients.

It is doubtful that, as a group, building industry people influence the award of many significant commissions. However, their image of a professional can affect a practice in other important ways. The quality (and quantity) of construction bids the firm's work attracts will be influenced by how valid the industry believes its estimates are. How specifications are written can determine how well the bidders understand what the firm really wants. How the firm processes change orders and accepts responsibility for its own errors and omissions will greatly influence the cooperation it receives in the field.

Ninety percent of a professional's communication with the building industry occurs in the normal progress of a project and is not subject to special communication programs. The important thing to remember is that the building industry is a large audience, it has its own highly developed grapevine, and the professional's standing may be significantly affected as the firm's image travels that grapevine.

THE PROFESSION

There is no greater measure of accomplishment for an architect or engineer than the recognition and esteem accorded by fellow professionals. This is reason enough to devote considerable attention to communicating with professional colleagues. In addition, however, one's standing in the professional community also has a good deal to do with recruitment of professional staff and can have a direct influence on the award of many commissions. Professionals are, or should be, represented on the boards of virtually every major institution and corporation, and as they disqualify themselves from serving the organizations they represent, they have a major voice in who else gets the job. In this sense the professional world is probably the only one where one's competitors are also a major source of work.

Communicating with the professional audience is at one and the same time the easiest and most difficult communication assignment. It is easy because the intraprofessional grapevine in every community is a constantly active communication medium, and the gossip that travels along it will shape an image for every practitioner, whether they like it or not. When, through ignorance, envy, petty jealousy (and sometimes malice) a professional's image with colleagues becomes seriously at odds with what it should be, effecting a change can be the most difficult communication job of all.

The major avenue for intraprofessional communication is through activity in local and national professional organizations. Fortunately for the professions, the spirit of public service is widespread among their members, and many of the most able practitioners devote a good proportion of their time to ACEC, AIA, AIP, ASCE, ASLA, NSPE and similar activities. This exposure brings recognition and also often demonstrates leadership abilities. It can pay off in more than personal satisfaction.

A professional whose reputation with colleagues is not what he or she wants it to be has a greater challenge. The professional grapevine, while always catty and often petty, is also usually astute. Professionals who are worried about their image within their own profession ought perhaps to look seriously at themselves.

PROSPECTIVE CLIENTS

Communication with those who may be in need of professional services is, of course, the name of the game. In one way and another, every professional's primary communication efforts will always be directed toward obtaining future commissions.

In evaluating prospective clients as a separate audience it is most important to recognize that what they want to know about an architect or engineer often has very little in common with the interests of the professional's other audiences. Prospective clients have their communication receiver tuned to a considerably different frequency.

First and foremost, the prospect always has a keen self-interest—they are concerned with protecting their own or someone else's money. Thus, they are very likely to concentrate on "mundane" aspects like cost controls, timetables, fees, and the like. Secondly, they are greatly influenced by how they think the choice of professional will be judged by others—

people are becoming overconscious of making the "right" decisions. This phenomenon will lead a prospect to avoid selecting a friend for professional services simply because the client does not want to be accused of favoritism. Finally, unlike some of the professional's other important audiences, the prospective client may be less concerned with visible quality of design. Even if the client is consciously seeking outstanding design ability, the selection may be made from other criteria because the client does not feel qualified to judge among the work of any half-dozen top-rank designers. Thus, the professional's graphic portfolio is, of itself, often a less important communication medium to a prospective client.

For all their uniqueness among the professional's other audiences, prospective clients are still normal, rational human beings. Communicating effectively with them is no greater an art than communicating with anyone else, once the differences are recognized.

These differences in point of view between the professional and the sales audience are very difficult for some to accept. We all know examples of brilliant young practitioners who receive lots of publicity, are hailed by their professional colleagues, and have excellent standing within the building industry, yet they have virtually no work. Why?

There can be many reasons, but in terms of communication one example is sufficient. Some practitioners teach in order to survive. And the longer they devote their efforts to communicating on an academic level, the less likely they are to put themselves across effectively to the lay client. (Involvement with the academic world is by no means a sure route to a professional oblivion—quite the opposite is often true—but the outstanding exceptions are those who became successful practitioners first, and educators second.)

Whatever the cause, it is generally true that those professionals who frequently have communication troubles with their clients and prospective clients, in one way or another spend too much time talking shop. They thereby tend to forget the prospective client is a complete novice to the vernacular of professional concerns. It is probably no accident that professionals who are most successful at getting their message across are often very involved in lay and civic activities. One wonders which is "cause" and which is "effect."

THE GENERAL PUBLIC

Discussion of the general public as an audience has been left for last because it is, truly, least important to the individual practitioner. It is probable that only a small fraction of the general public can even name a living architect or engineer.

The primary responsibility for communicating with the general public belongs to the professional societies. It is far more important that average citizens understand why society needs architecture and engineering than that they can name any single practitioner.

The general public that is of concern to the individual practitioner is the sum of those audiences previously discussed—clients, staff, professional colleagues, the building industry, and those opinion leaders in the community who may one day award or influence the award of a commission. For the average practitioner in a city of half a million people this total audience may be, at most, 500 or 1000 individuals. Many successful practices are conducted by professionals whose total audience is no more than 100 or 200.

The relative size of the professional firm's audience has a lot to do with the communication systems which will best serve its marketing needs. This is the subject of the next chapter.

6

UNDERSTANDING
THE MEDIA

"PR is part of marketing. It's the best advertising we have, only instead of billboards, we have people."
Marketer William Flaig

Equally important with understanding the variety among audiences is having an awareness of the difference between the media and techniques available for communication.

The human's faculties for perception of information are only beginning to be seriously studied (and it may be just as well if all the secrets are never unlocked). Marshall McLuhan compounded the problem by pointing out that the advent of new media (particularly television) effected actual and distinct changes in the perceptive faculties of new generations. At the very least, this means that what was true about communicating with people a generation ago may no longer be true today.

For our purposes, however, it is enough to accept these postulates:

No two people receive information in quite the same way;

No two people who receive identical information can be counted upon to put it to the same use.

Thus, no communicator can ever be certain that a given message will get through to a given audience. But the probabilities in each situation can be raised if one is selective about the media used.

For example, the mass media are best adapted to communication with mass audiences, and the degree they are used will depend on many circumstances. Advertising on television is probably the broadest mass media technique available today, and it is used extensively for communication of mass-sale consumer items.

Sometimes, mass techniques and mass media are used only for the projection of images (brand names), while selective techniques and media are used to close sales. Thus Ford Motor Company uses TV advertising to promote the image of the car, but turns to local newspapers to advertise prices—the consideration upon which final buying decisions are often made.

As one turns to more selective products and services, which appeal to narrower audiences, techniques, and media become similarly refined. A good illustration is magazines: Bacon's Publicity Checker, the standard reference index of magazines, lists 117 "General Interest" magazines ranging from *The Atlantic Monthly* to the *Saturday Evening Post.* In the same volume are 4,023 trade and professional magazines in categories ranging from Architecture to Warehousing with titles from *American Pigeon Journal* to *Tobacco Reporter* to *Medical Laboratory Observer* and *Wyoming Beverage Analyst.*

The primary reason for this refinement is simple economics: There are some 60,000 registered architects in the country, and *Architectural Record* goes to 56,700 of them. Its advertising page rate is $5700 or about 10¢ per architect reader. *Readers Digest,* on the other hand, has a total circulation of 17,750,000. Its page rate is $97,700, a cost per reader of only ½¢. However, even if every architect in the country reads it, the cost of a page ad in *Readers Digest* would be about $1.62 per architect. Obviously, it is much less costly, and more effective, to reach architects through their trade journals.

These are only the crudest numbers to illustrate the point. In the magazine business it is also important to know how many of the subscribers may *see* a given page of advertising. *Architectural Record,* with 200 pages of advertising per issue may have a significantly different "page readership" than *Engineering News Record* with only 50 pages of ads per issue. Then the question is how many will *read* the ad and *remember* it. In the world of publishing and advertising, fames and fortunes are made and lost over points like these.

Fortunately, the professional's selection of media does not rest on points as fine as whether to buy prime TV time at 8 P.M. or 10 P.M. Nevertheless, the choice will be motivated by exactly the same criteria—effectiveness to do the job. For comparative purposes, it is possible to classify the media available to a professional into three broad groups, according to probable effectiveness:

General media—where there is relatively little control of the audience and the chances of effective communication are probably no better than 1 in 1,000 or higher.

Selective media—where the audience is fairly well defined and probabilities may range from one in five to one in fifty.

Specific media—where there is no doubt about who you are talking to and the probability of constructive communication is as high as one in one, two, or three.

The probability comparisons given are merely illustrative, and are not based on any scientific study. The relative difference should be remembered, however, in considering individual communication techniques and media. Examples of each are described below.

GENERAL MEDIA

Newspapers

The great advantage of newspapers is that they reach a mass audience within a concentrated geographic area. This is of tremendous importance to department stores, those selling groceries, and any other advertisers (including design professionals) who want to reach that audience. In a publicity context, newspapers consume and publish a great volume of information, and it is not difficult to collect clippings if one works at it. The question about newspaper advertising and publicity is: Who reads it who is important to you?

In terms of absolute audience, the odds of reaching important audiences via general newspapers are very low. In a city of 500,000, the major newspaper may circulate about 100,000 copies. The audience the professional wants to reach may be, at most, 500 to 1,000. If one in ten of these sees the design firm's name in a given newspaper article, it will be doing well. Thus, newspaper publicity will give the firm a net useful audience of 50 to 100 persons. There are often other, more direct ways to reach such a small group.

The validity of newspaper promotion will vary for every professional firm depending on the nature of the practice and the size of the community in which it operates. Generally, the smaller the community, the greater the value of the local newspaper publicity. An article in the York (PA) *Gazette* mentioning the engineer of the new subdivision may be of value in helping the firm get another job. On the other hand, a similar article in the *New York Times* crediting the architect of a new school in the Bronx may be of no particular value at all. In general, the professional's prospective clients are much too selective an audience to be significantly influenced by routine newspaper advertising or publicity. Even if prospective clients do read newspaper articles mentioning a professional, the articles are written for a mass audience and they seldom contain information important to the client. One major exception would be a feature article about a truly innovative or new subject. Such a feature may bring several direct inquiries to the professional.

Television and Radio

The numbers game mentioned above for newspapers applies tenfold to radio and television. The broadcast media, today, have by far the largest total audience—but selectivity is the lowest of all. This is why most broadcast advertising is limited to brand names and mass-sale products.

As of this writing, few professional firms have found cost effective ways to use broadcast advertising, although there needs to be much more experimentation before it is ruled out. Publicity on television and radio is generally limited to news broadcasts, and because of the nature and pace of broadcast news, such publicity is of very limited value. Professionals find themselves in greater demand as panelists for "public service" programs (to which the radio and television stations are required by the FCC to devote a certain percentage of time). Many public service programs are very seriously conceived and could produce valuable communication for a professional if it were not for their doubtful audience. TV stations tend to broadcast "free" programs early in the morning, late at night, and on weekends at hours when the audience is so limited the time cannot be sold to a commercial sponsor.

General Magazines

There are only a handful of general magazines (*Readers's Digest,* etc.) left in circulation, largely because television competition has cut deeply into their advertising revenue. Perhaps the classic illustration is the *Saturday Evening Post,* which had its highest circulation in history in 1960 when it began to lose money. The problem is that the general magazine has an audience similar to television, but TV reaches that audience far more cheaply per thousand people than most magazines.

Business Press

For the professional, the one area of general media that can be of prime interest is the business press, particularly *Fortune, Forbes, Business Week,* etc., and the business sections of *Time, Newsweek,* et al. If the professional is in a position to obtain it, publicity in these magazines can produce very valuable returns. However, because there are so few of these magazines, they have their pick of the very best stories in the nation, and thus are likely to get their most publishable news and features from the 50 or 100 top firms in the professions. When they do publish a professional's work, the value is not so much in what is said as in the implied endorsement which comes from having one's work selected for publication. Such value can be enormous, but those who aspire to it must be realistic about their chances unless they are really doing breakthrough work in a significant market.

A number of professional firms are experimenting with advertising in the national business press for the same reason. A full page in some regional editions of *Business Week* costs about $6500, and may well provide that value in name recognition for the firm involved. Nevertheless, only a handful of the largest regional firms are in a position to capitalize on such an investment. Smaller firms can do the same thing at less cost through more selective media.

SELECTIVE MEDIA

By definition, the selective media are those which reach a fairly defined audience—either by the nature of a group or class of people, or by design, such as in the preparation of a selective mailing list. You know exactly with whom you are communicating, and such media are therefore much more to the point. There are four kinds of selective media which are of particular interest to design professionals.

Direct Mail

Mass mailings of the type that stuff business mail baskets have given the technique of direct mail a somewhat negative image. However, direct mail is an excellent selective medium because the sender has so much control over who will receive it. Experience indicates that even though an average 50% of the addressees on a mass mailing may toss the material in the wastebasket sight unseen, the other half will read and be exposed to the message. In commercial direct mail, for example, selling newsletter subscriptions or technical books to engineers or architects, a 2 to 4 percent return is considered excellent, and makes the entire mailing profitable. Building industry producers who use direct mail to reach architects and engineers with trade and product literature probably experience exposure as high as 20 to 30 percent on a regular basis.

Thus, direct mail is a very effective medium for architects and engineers to get their message in front of the specific audiences they want to reach. Mailings to selected lists of potential clients can pave the way for better firm recognition when it comes time to make a sales call or presentation. And mailings to past clients and other contacts personally known to the professional can achieve exposure rates of 80% to 100% and become the most cost effective single method of maintaining your image in front of those who know you best. (In this respect, the former ethical rule of the AIA that permitted mailings only to those with whom an architect had had personal contact, was really a license to use the most effective form of direct mail.)

A professional can easily build in a few years a mailing list of 500 to 1,000 important contacts, and by carefully planned use of direct mail, can communicate a great deal of information to this audience. Specific techniques and examples of how and when to use Direct Mail are discussed in Chapter 19. Methods for building promotion contact lists are covered in Chapter 8.

Professional Brochures

A well done and carefully distributed professional brochure can be an effective communication device to selective audiences. The relative effectiveness of a brochure is highly variable, however, depending on many factors of content and presentation as will be discussed in Chapter 17. Also, the distribution of a brochure is necessarily limited by practical considerations—the largest firm probably distributes no more than 200 or 300 a year after an initial mailing to whatever general mailing list the firm maintains. Thus, while its effectiveness may be high, the total number of people reached is rather small.

Speeches

The professional is always in demand as a speaker before trade and civic audiences, and a good talk can project a great deal of information. The limitation here is that the professional has only partial control over the makeup of the audience. One can decline invitations to speak before audiences that are unimportant, but it is not easy to create opportunities to talk before audiences that are important. A balanced professional communication program would include at least one or two speeches a year.

Trade Magazines

Trade magazines are another useful channel of advertising and publicity. This is because trade magazines as a group prize very highly the selectivity of their audience. An architect interested in doing medical facilities, therefore, can direct his publicity or advertising to *Hospitals* or *Hospital Progress* and know that the distribution of the first will reach 78,000 administrators in the American Hospital Association and that the second will go exclusively to 14,000 Catholic hospital officials. An engineer interested in doing waste treatment plants can reach 27,000 public works officials through the *Journal of The American Water Works Association*. There are trade magazines directed at every conceivable audience a professional might be interested in serving, and through the Audit Bureau of Circulation and Business Publication Audit services, their circulation claims are validated and published. No matter how selective the distribution, however, the actual readership is what will determine whether the professional's publicity or advertising produces worthwhile communication. Some trade magazines reach their audience through "controlled circulation," meaning they are distributed free to all members of a certain trade classification—such as school administrators. Other trade magazines have only paid circulation, indicating that the readers want the publication enough to buy it. It may be assumed that readership of magazines in the latter category is higher than among free circulation publications (although there are notable exceptions in certain fields).

A number of design firms are beginning to experiment with trade magazine advertising, especially where package dealers and other competitors are also advertising, as in the hospital field. At present, however, the greatest value of trade magazines to professionals is as outlets for publicity. Trade journal articles not only can reach a highly selective audience, but reprints of such articles can serve double duty as direct mail pieces and sales tools with proposals.

Of great importance in measuring the value of trade magazine publicity is the position and length of the article that may be published. The cover story of the best-read magazine will have a readership many, many times greater than a small news item appearing among the back pages of the same issue. All these considerations make the preparation and placement of publicity in trade magazines a highly refined process, and one which is best assigned to trained public relations people, as is discussed fully in Chapter 18.

Professional Magazines

By definition, a professional journal is normally an official publication of a professional society, such as the *AIA Journal* and the *Journal of the AIP.* For communication purposes, however, this category may also include the trade magazines aimed at the same professional audiences, e.g., *Architectural Record, Progressive Architecture, Engineering News Record, Consulting Engineer, Landscape,* etc. It must always be remembered that publicity in professional journals is read mostly by one's colleagues and competitors—not by prospective clients. Such publicity can be very valuable in recruiting a professional staff and building an image within one's profession. But an article about an architect's latest bank in *Architectural Record* will become a valuable development tool only if reprints are arranged and the professional mails or delivers them to bank prospects.

SPECIFIC MEDIA

Personal contact is the one and only certain communication medium available to the professional today. In face-to-face contact, one has the person's attention for a period of time. Whether meaningful information will get through depends only on the listener's interests and the persuasiveness of the presentation. This is the point to remember in understanding the media of communication. "Credit lines," clippings, brochures, etc. are all meaningless unless they communicate useful information to someone who, knowing it, can be of value to the professional. One can cite numerous examples of professional firms which have achieved considerable success without utilizing any medium of communication except one-to-one contact and word-of-mouth reputation. Conversely, there are examples of overly-aggressive firms which have invested in elaborate publicity campaigns and received no measurable returns.

SUMMARY

Because of the proliferation of information that bombards us all today, professionals will be doing themselves and society a service if they concentrate their communication program on the simplest and most specific media available to do the job. The message about the media for selling professional services is that personal contact and demonstrable merit will always account for 80% to 90% of the effort. All the other media can at best do only 10% to 20% of the job.

The Professional Business Development Process, which is described in the next section, is predicated upon the application of one-to-one effort in similar proportions. The message should be clear, therefore, that marketing is a job that cannot be turned over to some magical process, but, rather, requires the direct involvement of individuals working toe-to-toe with their prospects.

Can professionals sell their services?

You bet they can! And the more professionally and specifically they do it, the better returns they get.

II

The Business Development Process

There is a distinct step-by-step sequence by which professional services are offered, considered, and commissioned. While there is no practical or professional reason for a particular pattern to be followed, the sequence has become increasingly formalized, due in part to the activities of the professional societies, which are often asked to recommend procedures of selection; and also due to the policies of many public agencies, which must go through the motions of open and unbiased selection before awarding contracts for professional services. From the professional's standpoint this business development process can be divided into six steps:

1. *Market Research*—determining what kind of work you want to do and where to go to get it.

2. *List Building*—identifying those with whom you want to be in touch.

3. *Bird-dogging*—beating the bushes to find the live ones.

4. *Strategy Research*—learning enough about each quarry to determine your choice of selling weapons.

5. *Courting*—the art of making professional relationships.

6. *Paperwork*—preparing the forms and questionnaires.

7. *Interviews*—putting yourself across.

8. *Closing*—making the sale.

Not every commission requires all eight of these steps, but those steps which are involved tend to follow very much the same pattern and sequence from job to job.

This is the Business Development Process.

7

MARKET
RESEARCH

"If you don't understand the client's terminology, your credibility quickly goes down the tube."
                                                    Marketer Lyle Trease

The business world has given the term *market research* a dignity and mystery it does not fully deserve. Every professional who has ever set out to locate a possible commission has conducted a form of market research.

In reality, market research has two simple objectives:

1. Determining what kind of work you want to do.

2. Learning how to find those who sponsor this work.

The first element—determining the kind of work you want to do—is not as simple as it may sound. It is one thing to want to design a particular class of structures—say, hospitals or bridges—but this does not mean a market exists for such projects.

Ever since World War II the market for major building types has tended to fluctuate in distinct and recognizable cycles. The post World War II boom in family formation made land development for housing the number-one market for engineers, and the accompanying baby boom made schools the number-one building type for architects—first in elementary schools, and later in high schools. A decade and a half later the kids arrived at college and created a boom in higher education facilities. (Along the way, the Sputnik revolution caused a five-year rush to build new college science buildings, beginning about 1959.)

Government funding programs have had a profound influence on such things as hospital construction (the volume quadrupled from $200 million to $800 million per year around 1950 after passage of the Hill-Burton Act); highway construction (the ten-year Interstate highway system generated engineering fees alone of $1,640,000,000); waste water treatment (the environmental revolution of the 1970s, and associated federal funding schemes, created massive markets for sanitary engineers); and a host of other building markets.

These swings have been characterized by Harvard Business School Professor Theodore Levitt in a graphic way that can be a very helpful aid in looking at a firm's market prospects. (Fig 7-1.)

According to Levitt there are four phases in the life expectancy of products or services in the marketplace, and knowledge of their cycle can be critically important to the timing of marketing strategy. The phases, in design firm terms, are:

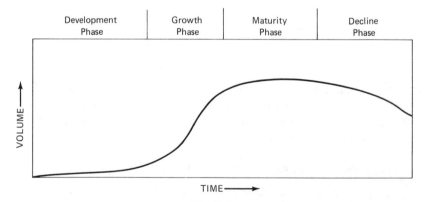

Fig. 7-1

1. *Market Development*—the phase when a new service is first brought to or required by the market, but before there is proven demand. Volume is low, and creeps ahead very slowly. There is generally very little competition. An example of this might be the handful of engineers and architects who saw the potential for energy conservation services some years before the Arab oil crisis. They knew they were correct about the long term need for conservation of nonrenewable resources, but they found few clients willing to pay fees to save still-cheap energy. Thus they had the jump on the market, but for a long time there was very little work to do.

2. *Market Growth*—the "take-off" stage when demand begins to accelerate and the size of the total market expands rapidly. In school buildings it was caused by a baby boom. In hazardous waste treatment services it was caused by the "superfund" legislation. Whatever the cause for the upturn in growth, this is the most advantageous time to be in a market from a marketing standpoint. At the same time there is the sharpest increase in demand and the fewest competitors. The energy prophets who waited through the lean market years had more work than they could handle when the Arab oil crisis hit and the market took off. Meanwhile, latecomers to the market had to scramble to get credible in order to share in the market.

3. *Market Maturity*—eventually demand begins to level off as the market need becomes satisfied and continuing volume is matched to national growth or replacement requirements. At this stage many firms have jumped on the bandwagon and there is more and more competition for a static market. The dilemma of space engineers who graduated just after the U.S. landed on the moon is a good example. As a general rule, by the time one sees examples of a new project type rising across the landscape (such as four-lane highways, or high rise office buildings) that market cycle is well into maturity and it is late to cash in big. The markets of the future are always the ones which can't yet be seen by every naked eye. In energy services, by the time the U.S. government began advertising in *Commerce Business Daily* for energy audits at all sorts of Federal installations, it was too late to get into the market unless one had prior experience or was associated with one of the established energy specialists.

4. *Market Decline*—eventually every service fulfills the demand or loses appeal or relevance. This is particularly true of the project and building type markets of interest to engineers and architects, because of our society's habit of "solving" national priorities one at a time, e.g., the interstate highway program, the renewal of hospitals, etc. At this mar-

ket stage there is more competition from firms that are geared up, but there is less and less work to go around. Firms that have not diversified into other growing markets generally decline. In energy conservation this stage occurred when the catch-up in existing structures was well underway and the market focused on energy conscious design of new structures. At this stage, the firms that had specialized wholly in energy retrofits found less and less to do.

Being sensitive to these cycles is of critical importance to business development planning in every design firm. Virtually everyone would like to know, in advance, what will be the next "big" market. There are also, of course, opportunities in fully mature markets for a firm that offers a significantly better service. Because no design firm ever controls even one percent of a given market, there is always room in a mature market for someone to capture a share of the work by providing something better than the competition. The techniques of market research discussed below can be used both to find emerging markets and to identify opportunities to provide a new or better service in existing markets.

How does one find the future markets that can't yet be seen? There are specialized market research organizations which make a business of analyzing and forecasting such trends. Their data is expensive ($10,000 and up) and is most helpful to firms which are trying to capture a percentage of a market on a regional or national basis. Only the largest professional firms are likely to find such definitive studies useful. The smaller firm can do its own market research with adequate accuracy on the local level.

At the broadest scale, effective market research can be done by reading the client and trade press with an open eye. Business pages of daily newspapers (which surprisingly few professionals bother to read) reveal a great deal if read regularly—particularly about plans for corporate activity and government policy that will affect building. The annual forecast issue of *Engineering News Record* (it comes out in January and is available in almost all libraries) gives detailed construction projections for every building type and geographic region and is an excellent index of the current market. In reading construction forecasts, however, it should be remembered that the lead time for architectural and engineering commissions is about 18 to 24 months before building begins.

Some professional firms subscribe to *Dodge Reports* (cost can run to from three to five figures annually, but the average is about $1,000) and use it to find current leads. But careful reading of the daily issues can provide insight into local trends when, for example, the first new apartment project appears and one has been reading nationally about the reviving rental housing market. *Dodge Reports* tries to note when professionals have not yet been selected for a project, thereby occasionally providing direct leads.

At the local level, however, professionals do not need to stuff their mail box or become economists or statisticians to conduct useful market research. The easiest method is to ask someone who already knows.

If one is interested in higher education, call on the facilities vice president of the local university and ask what the market will be for three to five years. Chances are the source person will be delighted to discuss the market— and the source might make a useful contact at the same time. If possible, call on two or three similar people at other colleges and average their opinions.

The same process works equally well with corporate facilities engineers, hospital administrators, public works officers, and almost any class of institutional client. In the real estate world, local commercial realtors and industrial realtors (there is an important difference) are willing prognosticators of their markets.

The process of conducting market research is just about that simple. The first step is to establish:

What market or project type(s) will be investigated?

What territory (local, statewide, national, etc.) will be studied?

Who are the sources that can be contacted for information?

There are knowledgeable people available in virtually every field who, if approached and asked, are glad to share with architects and engineers their view of their own market. One of the best sources to find them is trade associations, and an association exists for just about every field you could imagine—and many you can't. The bible of trade associations is Gale's *Encyclopedia of Associations* which is available in public libraries.

When you have identified the associations that represent people in the field you are researching, call or write their national office and find out who of their chapter or national officers are based in the territory you are studying. You may also want to interview some of the association staff people to learn what they know about the market, but never stop there. It is the volunteer officers and directors of most associations who are the real hands-on authorities in the field. They are hiring engineers and architects in their own organizations every day, and what they have to say about the market for professional services is derived from direct experience. Also, people who are active in their own trade association tend to be in close touch with their peers, and therefore can speak about the broad market with some authority.

Making contact with such people is not difficult. By virtue of their office in their trade association they tend to be more approachable. Just phone (or better still, write first) and say:

> I am contacting you at the suggestion of *(the association staff)* in your capacity as (officer) of the _____ association. I am conducting a market research study into the (need for, outlook for, etc.) such-and-such design services (or project types). As an authority in your field I believe you can be very helpful to me. I would like to make an appointment to come by and ask some questions.

Few people will not respond to the opportunity to be an authority in their own field. Experience demonstrates that almost all people approached in this manner welcome the contact and give freely of their time and knowledge.

Once you have identified a knowledgeable source, the key to getting successful market research data is knowing what questions to ask. It is advisable to prepare in advance a written questionnaire for the subject you are researching. The following are general directions most design firms would want to cover:

a) What is the volume of work anticipated for this project type for the next 12, 24, and 36 months?

b) Is the future trend of volume up, down, or stable from the recent past? Are there particular causes for these trends?

c) What is going to be the nature of typical projects in terms of program requirements, scope, etc.? Are there any particular current or recent projects you can cite as prototypes for the future?

d) What scope of professional services are normally required by clients for these projects?

e) At what stage in the planning process are these services normally engaged? Are there predesign consultants who may be involved in the preplanning?

f) What are typical professional services fees for this sort of work?

g) Is there any pattern to the process by which consultants are considered and selected for this work? Are any standard forms or formal paperwork usually required?

h) Do clients for this work generally seek consultants who are specialists, or are firms without prior direct experience given equal consideration? Are joint ventures or associations common?

i) Is there any way to judge the competition for this kind of work? How many firms are actively seeking the work? Can any of these be cited by name? What are their strengths and weaknesses?

j) How might one develop a list of organizations who are potential clients for this work?

k) Is there any way specific clients for this type of work can be identified before professional consultants are selected? Is this generally public or private knowledge? At what stage does it become known? Is there any regular source of leads or market information which can be contacted or subscribed to?

The answers to these kinds of questions can be invaluable in helping a design firm focus its marketing effort. Sources who can provide such answers, in addition to trade association people, include any active clients in the market, other consultants, trade publication editors, financial officers of banks who underwrite projects in the field, and, where they exist, economic development departments of state and local government, utilities, and Chambers of Commerce.

In practice, a good market research overview can be gained through as few as three to five interviews with such sources, provided their answers begin to establish a similar pattern. If there are wide differences in the opinions, obviously, further interviews may be needed to get a clear perspective.

It should be noted that this kind of market forecasting is equally important for established professional firms already serving a share of the market. If, for example, hospitals have been 25% of a firm's practice for several years, an annual diagnosis of the future hospital market can help avoid unhappy surprises when, for example, one of the federal funding programs suddenly expires.

So much for the first step in market research: The point is to decide what kind of work the firm would like to do, and then investigate how large the market may be and how long it is expected to last. When the answers are in hand, the professional is ideally equipped to begin the next steps in the business development process.

Accomplishing all the market research which a firm could do may sound like a frightening amount of work—particularly to a small organization where the principals are already up to their ears in practice. Don't be discouraged. A little effort is all that is needed to yield returns. If any professional firm ever actually did a full market research study on every market open to it, they would turn up more work than they could possibly grow fast enough to handle. Don't believe it? Consider this example: A West Coast architectural firm developed some recognized (published) experience in out-patient medical clinics and had referral projects in three states when they decided to do some market research into the emerging needs of Health Maintenance Organizations. The firm's lead finder, who was exceptionally good at information gathering by telephone (he was a former Dodge reporter), was assigned to research the market west of the Mississippi and find out what the opportunities might be for an out-of-state firm with expertise in the field. Within three weeks of phone calls he had established:

1) HMO's generally were on the verge of going into building programs in a big way;

2) There were no established specialist architects pursuing the field and the firm's published work was recognized as the prototype to follow;

3) HMO clients had no hesitation working with an out-of-state specialist, and would expect to pay extra to get that expertise;

4) Public agencies in each state were tracking HMO programs and could give names and addresses of active potential clients;

5) In the first 30 sample calls to HMO's on these lists, 17 potential clients interested in talking about architectural services were identified.

This is an unusual example, of course, but it is not unique as a demonstration of what effective market research can do. The information is out there for those who take the time and energy to ask.

from: **the Coxe Letter**

CASE STUDY #81: MARKET RESEARCH MAKES BIRD-DOGGING EFFICIENT

Four years ago Lloyd Bakan, Marketing Manager for the 130-person Spink Corporation in Sacramento, California, hired a freelance market researcher to help him switch from a shotgun approach to more efficient cold-calling. He asked the researcher, Randall Shores, a former county administrator, to identify all public works projects in a 14-county area that hadn't yet been awarded to consultants.

Shores got copies of the capital budgets and located (sometimes after as many as 12 phone calls) the person in each who was responsible for hiring consultants. He talked by phone to each one and identified all unawarded projects and the budget and timetable for each. ("The best time to call," says Shores, "is about three-quarters of the way through the fiscal year. It can vary from agency to agency, but after the first call you know when the right time is.'') The project list has been updated and expanded every year since.

"What it has allowed me to do," says Bakan, "is to walk into a City Public Works Director and say, 'I see you have $x budgeted for such-and-such project, and we're interested in being your consultant.' It's a lot more effective than just asking, 'Do you have any jobs?' Twice directors I talked to didn't even know the projects had been budgeted. It saves an enormous amount of my time. And it gives us a jump on the competition, because we know about the job before they do.''

To date Bakan can identify two commissions that started with the market research: an $8 M county courthouse, and a $400 M waste treatment facility.

8

LIST BUILDING

"If you don't know who you are going to call on next Monday or the week after, you've got problems."
Marketer Kevin G. Barbera

Once a firm has selected the markets it want to pursue, the next step is to identify those who commission professional services for this work. At this stage the objective is to prepare a list of likely prospects to be contacted later in search of leads. Developing a good list can involve a multitude of procedures depending on the building or project type, the geographic area, and the nature of the client group, but it is nowhere near as difficult a task as some firms perceive it.

A young firm interested in doing medical work can probably list from off-hand knowledge the eight major hospitals in its community. If the firm has a regional or national interest in doing hospitals the research problem becomes more complex. In such cases it is often best to turn to the trade association of the market—in this case the American Hospital Association. Through its organized state and national activities and publications, the AHA is an excellent source of information to classify prospective hospital clients because it can provide information on size and growth of member institutions, trends in their facilities needs, etc. (AHA also welcomes participation by professionals. The AHA national convention each year draws hundreds of design firm registrants.)

The same process can apply to some seemingly quite obscure markets. There is a national academic group that keeps its finger on what colleges are sponsoring new theaters. If one should be interested in public arenas, there is an International Association of Auditorium Managers which publishes an annual directory listing the status of every proposed auditorium/convention center in the country—many of them well before professionals have been selected.

For federal government work, the major professional societies (e.g., ACEC, AIA, etc.) often publish directories of public agencies contracting for building design, giving the names and addresses of Government departments which purchase professional services.

Probably the most difficult markets to follow are in business and industrial fields, but a lot can be learned about these simply by knocking on doors.

For example, a few years ago a young architect in Philadelphia chose to explore the business and industrial market—probably the hardest of all to effectively research—because he wanted to diversify a largely institutional practice. After learning all he could about available information sources, he decided to focus on a local publication called the *Greater Philadelphia Industrial Directory*. This volume lists every company in the metropolitan area with 25 or more employees or annual sales over $100,000. The architect sifted the 3,000 companies in this directory and developed criteria (e.g., type of business, number of employees, etc.) to help him evaluate the probable potential of different companies. Using this base, he listed what he believed would be the 100 most likely to have use for his services. This ended his list building.

Then he began bird-dogging the companies on the list. The fourth company he called upon was a local utility. The Vice-President for Facilities gave the architect a very cordial reception, but at the conclusion of the meeting expressed doubt that the company would have any new work for some time, and candidly noted that the company already had available to it an extensive list of experienced architectural and engineering firms. They parted cordially, but the architect crossed the company off his list.

Exactly two weeks later the company called the architect and commissioned him to design a $500,000 warehouse.

What happened? Evidently the warehouse project was as much a surprise to the Facilities Vice President as it was to the architect. However, when faced with the sudden need to find a professional for the relatively modest job, which had to be designed on a fast timetable, the client apparently feared the job might get lost in one of the big A / E firms with whom the utility was accustomed to dealing. So the last professional who had come by asking for work got the job.

Very often the process of finding new business is just as simple as that.

It must be emphasized that the particular channels to the market vary with every situation, but the important point is that the starting place is to have a contact list of possible sources of work.

A list of national directories which can be helpful in some of the most common markets is included in the Appendix. But the author emphasizes, as the example above demonstrates, that the most valuable references are usually local or regional. Chambers of Commerce, state and local industrial development agencies, local offices of the federal Small Business Administration, and public libraries are the places to start. Then follow your nose.

The key to being successful at active marketing is being certain that every person with marketing responsibility in the firm has a list and knows what is expected of them with respect to that list.

In a typical firm, those lists can vary considerably in nature and length according to the individual marketing role. Following are some hypothetical but not unrealistic examples.

The firm's Principal Architect may have a "community list" of 16 important sponsors of work in the area. The list may range across many project type markets, but represents the local client "establishment" and are the type of people whom the principal, as the firm's most visible and active community leader, can approach. The Principal Architect's assignment over the coming year is to have at least one face-to-face discussion about the firm's services with each of the 16 organizations on the list.

The firm's Principal Engineer, a specialist in structures, has been assigned to help the firm break into the industrial plant market. A list of 30 growing industrial companies has been developed with data obtained from the State Industrial Development Agency (which is also on the engineer's contact list). The Principal Engineer's assignment for the current year is to establish contact with a facilities manager in each of the 30 companies (a rate of less than one a week) and find out which might be interested in the firm's services. This will yield a shorter list for regular recontact in the future.

The firm's Project Manager in charge of hospital work has a list of eight past hospital clients of the firm and is responsible for knowing what work each of them has coming up.

Meanwhile, the firm's Marketing Representative has a list of 20 state environmental agencies dealing with hazardous waste and is assigned over the next three months to contact them and identify the names of all multisite private waste disposal companies in the 20 states. Then the Marketing Representative will contact those private companies to find out which ones may be opening new sites and might need the firm's services in writing Environmental Impact Statements for hazardous waste disposal facilities.

When list building in a design firm is refined to this degree, no one in the firm need ever ask on Monday morning: "What marketing should I be doing this week?" And when every person in a firm with marketing responsibility knows the answer to that question, in advance, the firm is well on the way to developing new business.

from: **the Coxe Letter**

CASE STUDY #84: MAILING LIST GROWS FOURFOLD IN A YEAR

A year and a half ago, the Peoria, Illinois, A/E firm Lankton Ziegele Terry and Associates started an internal analysis of its markets, its clients, and its goals. A young MBA candidate, looking for summer employment, was hired to examine the details of the firm's operations, its history, and profits and fees in various project types. "He compiled a data base that helped us make decisions," says Vice President of Business Development Herman Thun. "It gave our goals and aspirations some structure."

"One thing that became apparent was that the number and type of people we were talking to on a regular basis was limited," says Thun.

At the time the firm's mailing list consisted of 150 names. "A high percentage of our work had been repeat clients, and we wanted to have greater versatility and more diverse markets in the future," notes Thun. The firm set a goal of 1000 names within 3 years.

Systematic list-building was initiated, and when the cold-calls—either in person or by phone—are made, the firm always asks the contact if he/she may be added to the mailing list. "We haven't had a refusal yet. Prospective clients seem eager to know about what architects are doing."

In addition, the "past clients" on the mailing list were reviewed. Generally each project was represented by only one name. Thun added the names of all the decision-makers in each client group, "including the people in the lower strata—they often influence decisions, and eventually they're likely to move up in the organization."

Another source for names is interviews. Even when the firm loses a commission, everyone on the selection committee (plus other critical members of the client group) are asked if they might be put on the mailing list. "Again, so far no one has declined," says Thun.

After one year, the mailing list has grown to nearly 600 names. In the meantime the firm has initiated a periodic newsletter. The results: wider name-recognition for the firm through its direct mail campaign, and the firm has an organized system of making new business contacts to a broader audience.

"But the best thing," Thun muses, "is what has happened internally to us. We've started a process of internal dialogue. We are more aware of how we reflect ourselves to others. Our office walls are now literally mirrors. We realize we have to think about how the people out there perceive us—and we have to do more reaching out."

9

BIRD-DOGGING —THE ART OF TRACKING THEM DOWN

"You can't possibly do it by staying in the office."
Marketer Norton C. (Sandy) Wright

The sales activity known as "bird-dogging" is just what the term implies—finding and pointing.

The term as it has come to be used in professional firms was coined by nonprofessionals who sometimes do the job on the payroll of architect/engineer organizations. It implies no disrespect or condescension: Rather, it is a clear recognition of the difference in role between the one who seeks and finds an opportunity as contrasted with the one who sells and closes the deal.

The bird dogs in professional firms know that only a full-fledged professional can effectively appear at an interview and sell a service. But they also know that the job of finding that interview opportunity is a time-consuming and specialized contact activity which can be done by any trained person.

And there is no reason to be shy about engaging in bird-dogging. Just as the untried market looks awesome to the professional, so also does the process of selecting professionals completely awe many prospective clients. Such clients are just as anxious to be found as the professional is to find them. Unless the prospective client is offered an alternative, the client will often select a professional simply by asking for a reference from someone who has done it before.

It is the author's observation that probably 40 to 50 percent of all new commissions are awarded by referral—many with no competition at all. This is very nice for those receiving the referrals, but it does not necessarily represent the best process for establishing an agency relationship or to secure the best architect or engineer.

The viewpoint of the client is the factor most often overlooked by those professionals who decry enterprise bird-dogging and selling. It would be very nice, of course, if every client selected architects, engineers, planners, etc. with the care and concern of the patron of old. But today's client is the antithesis of this archetype: Often the client considers the building more like a commodity within the broad spectrum of the organization's activities; the client has little or no understanding for the role of a professional as different from other vendors in the building process; and the client has no background from which to evaluate the qualifications of different professionals to do his work.

In this climate, the burden of assisting the client in choosing the right professional falls clearly on the professional. How many times has one looked at a structure and said: "Why didn't they do it better?" Those who ask this question generally intend to blame the client. Actually, it is often the professionals who are at fault because the client was not offered an alternative and hired the first architect or engineer to be found.

This is the social argument in favor of bird-dogging. On the professional side, it follows that those who would like to see clients adopt higher standards must accept responsibility for educating them. Some of this can be done via public relations by the professional societies but in the main it requires one-to-one contact. Ergo, if the "good guys" fail to seek opportunities to preach their gospel to the decision-makers, what right have they to complain when the "bad guys" get the work?

As a general rule these days, very few responsible people react to unsolicited information. Thus, a written request for consideration seldom yields results. At the same time, interestingly enough, most of these people are quite willing to talk. Thus, the professional whose first approach is to appear in person will usually be far ahead of the professional who comes in by mail.

Why then do professionals continue to resist bird-dogging as a valid and necessary part of a selling program? A possible answer lies in a frequent misunderstanding of the purpose of the function. If the bird-dog considers him or herself as a salesperson, the work can be very discouraging because a high proportion of contact calls do not result in a sale. It is human nature to avoid rejection, and only people with the toughest hides can handle slammed doors like the door-to-door peddler. Few professionals are that tough, because the design professions tend to attract people whose behavioral profile is more support-giving and nurturing and much less aggressive and risk-taking than the classic hustler.

This alone would justify limited bird-dogging by design professionals if it were, in fact, a sales function. But bird-dogging by definition is intelligence gathering, not sales. The bird dog seeks information for the hunter, and when the dog finds a target it only points the direction for the hunter to aim.

The bird-dogging of potential clients for a design firm is exactly the same. The job is to find information, not to sell. When this is understood it becomes much easier to make cold calls because in a real sense rejection is not an issue. It can be just as valuable to knock on a door and have it slammed in your face as it is to receive a warm but useless reception. In both cases the bird-dog has learned not to bother with those two doors again. This is useful intelligence because it helps narrow the list to those with whom the firm has a realistic chance of dealing.

Thus, in practice, it cannot be emphasized too strongly that bird-dogging calls should not be used as sales calls. If one is asked during a bird-dog call to make a sales pitch, a very proper—and effective—answer can be:

> I really didn't come here today to sell our services. I came just to find out if you may ever need the kind of services we offer. If you would like to know more about us I will be very glad to go back to my office and prepare a proper presentation. Perhaps you would even like to come to the office to see it. Now, tell me what work you have in mind so we can put together a proper response . . .

In the same context, many experienced bird-dogs do not carry brochures when they make calls. This allows a request for a brochure to be answered later, by mail or messenger or return visit, accompanied by a tailored sales letter aimed at the particular opportunity that has been uncovered.

But this is all getting ahead of the game—in order to reassure the doubtful that the effort at bird-dogging, when seen as intelligence gathering, can be a comfortable function and should be worthwhile.

The previous chapters discussed the importance of deciding what markets one wants to pursue, and the need to identify those who sponsor this type of work.

Bird-dogging, then, begins with the list, which, ideally, contains:

1. The names of the organizations, institutions, or companies in a given territory that may use the particular service one is interested in providing.

2. An evaluation of the list in order of priority for contact.

With this list in hand, the object is to learn:

3. The name and address of the executive in each organization who considers professionals and receives proposals.

And from these contacts the professional will want to find out:

4. Whether the organization uses professional services such as the firm offers; and

5. Are any specific projects in prospect in the foreseeable future?

Each of these steps requires a particular strategy in order to make the bird-dogging effort both efficient and productive.

GETTING THROUGH

Frequently, the first question in bird-dogging is how to know who is the right person to contact in the target organization. How to get this desired information will depend on the situation. If it is a public or quasi-public institution, a telephone call may suffice. For example, if the local County Manager can be reached by phone, one could say:

> I have an (engineering or architectural) office in _____ and I am interested in learning more about the North County District. Can you tell me how the district evaluates professional services and whom I might contact in order to learn more about your building plans?

To be sure, some people will brush off such a call. But many will be genuinely helpful, and these are most likely to be the ones worth following up. Depending on the answer to the initial question, one or all of the following additional questions should be asked:

> Should I contact _____ by letter, or would he or she see me if I called for an appointment?

Do you have a standard questionnaire I should fill out?

Do you know if there are any future building programs currently under consideration? (At this stage, an unexpected answer to this question could set the priority for your response.)

When a phone call won't do, sometimes a letter will work, particularly with large businesses or corporations. For example, this approach is entirely proper:

President (By name)
XYZ, Incorporated
Dear Mr. _____

Our firm has diversified engineering capabilities which we feel may be of interest to XYZ, Incorporated. Could you tell me if your organization ever uses outside consultants for such services and, if so, whom I might contact in order to learn more about how you manage for your facilities needs?
Sincerely,

ABC Engineers

Few presidents will answer such a letter in person, but few reputable organizations will ignore it either. Usually, the person who answers is the one the professional will want to meet. If it is not appropriate or practical to learn what you want to know by phone or mail, then make a personal call. This is particularly useful with large, complex organizations such as government agencies, and with very small outfits, such as home builders. The first visit may only get as far as a succession of secretaries, but if questions are asked in a sincere and friendly manner, they will get lots of honest answers. And if one finds the right secretary, later calls or letters for an appointment are likely to get right through.

In many organizations, the easiest way to get the right information about who to talk with is to call the switchboard and ask for the president's secretary. Then tell that secretary your purpose and ask:

Is this something I should discuss directly with the president or is there someone else in the organization whom I might see?

Generally the staff of the president is responsible for relaying away as many incoming requests as possible, but only to next level down in the organization. (This is quite different from the perspective of the switchboard operator who will probably refer you to the lowest level where you may get your answers.) When the president's secretary tells you that Vice President Smith is in charge of facilities matters, it is perfectly proper—and invariably effective—to write or call Smith and say:

The President's office referred me to you and I'd like to make an appointment.

If you doubt that this opens many doors, just ask the salespeople who get through to you how they got there!

Asking the Right Questions

Once in touch (by phone or face to face) with the person who knows the answers you seek, it is your job to get them by asking the right questions. Going directly for pay dirt—e.g., "Do you have any jobs today?"—is rarely effective. It is better to have in mind a picture of the background you would like to know about the organization and go after it first with questions like these:

1. Can you tell me what kinds of facilities your organization uses?

2. How are they generally obtained (e.g., Do you build, buy, lease, etc.)?

3. How far ahead do you plan for new or expanded facilities?

4. How is your company (department) organized to manage the facilities effort?

5. Do you use professional consultants in designing these facilities?

6. Are there particular qualifications which you look for when you select consultants?

7. Are there any future projects currently in the works for which you may need consultants?

8. What would it take to be considered for this project?

Obviously, the specific questions in each approach must be tailored to the situation, the relative sophistication of the individual you are talking to, and the chemistry of the encounter. But the point of this chapter will have been made if it is clear that the best approach to bird-dogging is to make each contact with the attitude:

We want to learn if we should be selling our services to this organization. What can we find out that will help us make that judgment?

Learning the Rules

The importance of learning during bird-dogging what a professional must do to be considered for an assignment is crucial to the success of any aggressive business development program.

The author knows of firms who have developed lists of potential clients and then, with no further research, have set out to offer their services by personal calls or by mail. The results have been uniformly discouraging.

If the way to be considered at XYZ Corporation would be to take the chief engineer nightclubbing, then don't bother writing letters to the president. If the way to be considered for work in North County would be to offer five percent of the fee to the county political chairman, then calling on the county manager will be a waste of time.

The author by no means suggests that a professional should do what is proposed in these cases. Most professional firms would not want to have anything to do with such clients. This is just the point.

If you know what it really takes to be considered, you can save a lot of time and effort. How do you learn the facts? Ask.

A not untypical case illustrating the value of asking involves the experience of a new A/E firm organized a few years ago by separate engineering and architectural firms that did not want to merge but wanted to market the advantages of multidiscipline services. They therefore organized between themselves a new firm and each contributed a strong principal to it. The architect and the engineer assigned to the new firm then faced the challenge of developing a marketing program.

Their first decision was to limit their market to the large metropolitan area where they were located. Then, within this area they developed a list of 50 potential client organizations, about evenly divided between public agencies, corporate, and institutional clients.

After three months, during which the two principals made a discipline of making four contacts a week, they reported the following results:

In 46 of the 50 organizations they were successful in making appointments to see the person they wanted to see—usually the chief executive or the facilities officer. (This demonstrates clearly that approaches by design professionals are welcomed by nine out of ten prospects.)

After discussion, 37 of the contacts said they would consider this newly organized A/E firm for assignments when and if they had work. (This provided the professionals with a reduced list of prime prospects for continuing recontact in the future.)

Nine of the contacts revealed particular projects for which A/E's had not been selected, and for which the firm would be considered. (Not all of these live leads were immediate, but it gave the professionals lots of targets toward which to gear up their courting, strategy research, and presentation systems. (See subsequent chapters.))

Two prospects actually selected the firm for projects within the 3 month period. Both of these were small planning assignments (one, an energy audit for an industry, the other a rehabilitation study for a section of a hospital) but they gave the firm a track record to publicize with other prospects and a foot in the door toward later work with the particular clients.

These kind of results from bird-dogging are not unusual, particularly when dealing with a prime list of recognized potential clients. The yield is likely to get thinner as the list expands to include lesser known organizations. But the principles are the same.

The writer, from experience, would offer this rule of thumb: If a proper job of market research has been done, every six to ten bird-dogging calls should turn up one live prospect. (A prospect is not necessarily a commission, as will be discussed in the next chapter, but the purpose of bird-dogging is only to locate prospects.)

Using the one-in-six/ten ratio, a busy firm can make one call a week and hope to turn up a prospect every two to three months. A more hungry firm may make an average of one contact a day, and yield 24 to 36 good new prospects a year. If, after a genuine effort has been made, prospects are not being turned up in these ratios, it is probable that the potential of the market has been misread, or the bird-dogging is not getting to the right people. (See Measuring the Marketing Effort, Chapter 24.)

RECORDING THE INTELLIGENCE

Every bird-dog contact should result in some kind of record for future use by the firm. There are as many different formats for so-called call reports as there are marketing directors, and it really doesn't matter which approach is used. The author's observation is that most live marketing data is transmitted verbally in design firms, and there is no particular value to dictating and filing lengthy contact memos unless it suits the personal style of those involved. There are, however, a few records which should be made from every contact because the information will be invaluable later. Any call report system that records these things does the job:

1. Proper name, title, address, and telephone for the person and organization contacted. (It is amazing how many follow-up sales letters are written to incomplete or misspelled names simply because the bird-dog neglected to ask the contact's secretary for a business card or letterhead.)

2. A *yes* or *no* decision on whether this contact should be added to the firm's mailing list. One can never reconstruct the list of these contacts a year later when the firm publishes a new brochure. The time to develop a marketing mailing list is every day that organized marketing is being done.

NEW BUSINESS CALL REPORT

Name: _____

Title: _____

Organization: _____

Mailing Address: _____

Telephone: _____

Date	Last Action Taken and/or Information Gathered
	Next Action:

Add to Development Mailing List: (   ) yes; (   ) no.

Fig. 9-1

3. The *next action* to be taken with this contact. Effective selling never leaves the initiative with the client. If it really is a "Don't call us, we'll call you" situation, you might as well remove the contact from your prospect list. An effective marketing program requires that everyone making contacts in the firm decide after each call what is the *next action.* It may only be "Recontact in 12 months"; or "Take new brochure as soon as it is published." Or it can be some immediate selling tactic. Whatever it is, the *next action* should be recorded on the master prospect list kept by the marketing coordinator, and keyed to the calendar. Then there is a way to follow up and see if the action is taken.

One format for a new business call report incorporating this information is illustrated in Figure 9-1.

LEAD FINDING SERVICES

The ultimate goal of bird-dogging, of course, is finding leads for future work, and there are other ways to do this than making all the contacts yourself. There are a number of lead services that can be subscribed to in different sections of the country that will in one fashion or another canvas the market and report leads. These services come in several varieties:

Newspaper clipping services. One can subscribe to a clipping service and ask for articles about future plans for any given type of project where no engineer or architect is mentioned. Clipping services are most useful in uncovering public work, especially in smaller communities where plans for any new projects are news. But they can also track very large projects like major bridges, airports, etc. Their drawback is that the clippings come in completely untested, and someone in the firm must call in each case to find out if consultants are in fact already involved.

Published "lead sheets." Dodge Reports, mentioned in the previous chapter, is probably the most widely known example of services which purport to publish names of projects for which no engineer or architect has been selected. These services can provide valid leads, but the information goes to a large number of subscribers who may decide to compete for the same commission. And many private clients, to avoid being besieged by design firms, go out of their way to conceal plans from these services.

Defined market lead services. From time to time lead services are offered in specific markets or project types (often a single state). And many of these limit the number of subscribers in an attempt to give some exclusivity to the information. These services can be very cost effective if the market they cover matches your marketing plan.

Contract lead finding services. There are also in certain areas contract services that will screen prospects for one architectural or engineering firm at a time and report which clients have projects in the planning stage.*

All these lead services can have value to those firms that are lead oriented in their marketing. The choice should be made on simple economics. In 1982, F. W. Dodge's survey of field sales calls in industry estimated the cost of a personal visit to a prospective customer at close to $200. It is probably even higher for design firm bird-dogs after all time and expenses are counted in. If it takes six to ten calls to turn up a worthwhile lead (see above), this equals $1200 to $2000 each. A lead service that may have a lot of chaff but yields a few good leads within that cost range may be a good buy.

In practice, many firms that have access to good lead services use both—direct bird-dogging in the specific markets where the firm wants to be known, and a lead service to find targets of opportunity elsewhere.

As mentioned in the previous chapters, there are always more opportunities out there than any one firm can even handle. This is why so many lead services come and go—they experience high turnover in their A/E clients because most firms can't use all the information they turn up.

The value of doing your own lead finding is that it puts you in control of the type and volume of leads you seek. This is all there is to bird-dogging—searching for desirable opportunities to sell the firm's services.

Bird-dogging ends when one stumbles onto an active, live opportunity.

A quarry is in sight. Now is the time to take aim.

from: **the Coxe Letter**

CASE STUDY #11: THE VALUE OF BIRD-DOGGING

Four years ago the then 16-man firm of Broome, Selig, Oringdulph & Partners, Portland, started a program of bird-dogging prospective clients.

---

*The names of active lead finding services are not listed in the Appendix of this book because there is frequent turnover among the firms that do it. However, a relatively current list of lead services can be obtained from the leading newsletter in the field, *A/E Marketing Journal*, P.O. Box 11316, Newington, CT 06111.

The first target was higher education facilities. Members of the firm contacted 25 colleges, universities, and junior colleges over a period of several months. Results at the time didn't turn up any actual work, but the firm received encouragement to follow-up for work when conditions got better.

A second effort focused on 58 industrial and commercial firms. These contacts produced information that most of the companies on the list would not be commissioning anything of size of interest to the architects. The list was thus narrowed to 15 companies found to merit continuing contact.

A third program was aimed at maintaining contact with school districts where the firm had worked in the past, and initiating contact with a dozen more.

The results to date: A law school addition, a fine arts building, and a science building for one college; projects in three school districts where the firm had never worked before; and a series of general commercial projects. The firm now has 32 people and is still growing.

"It made a believer of us," says partner John W. Broome, AIA, "I'm convinced we wouldn't have the work if we hadn't been knocking on the doors."

The bird-dogging work is divided between Broome and several of his partners, who make about 40% of the calls, and project manager William S. Findlay, who covers the rest. Together, Broome says, they average 4 to 6 calls a month, although it tends to be irregular: "Weeks will go by when we're busy and we don't do any calls; then we have a big flurry."

"It's a real long range thing," Broome emphasizes. "Many times we will call back three or four times and still find nothing. But we wouldn't have been considered for most of this work if we hadn't perservered."

"When you start doing this you hate it," he says, "but as you become more accomplished it becomes a challenge. Then you sort of get a kick out of it."

10

STRATEGY RESEARCH —FINDING THE RANGE BEFORE FIRING

"Chasing all the rainbows is an impossible task. If it is worth going after it deserves your best efforts."
Marketer Donnell V. Grimes

There is nothing in the process of selling quite so exciting—or intimidating—as the moment when one uncovers a genuine prospect who has an active project under way and is seeking the professional services one offers.

Often, this moment arrives by mail from one of the organizations the professional has been bird-dogging. The letter will outline a scope of service required, and will ask if the firm wants to be considered. A questionnaire may be attached, to be completed and returned by those who want to stay in the running. Letters like this usually go to somewhere between one and two dozen firms—often more.

Occasionally, a firm may get such a letter from an organization it has never heard of or contacted. Sometimes this comes because of a referral from a satisfied client, and if this is so the professional has already scored a few points.

At other times, the inquiry may come from a well-intended group which, in the name of impartiality, has sent the letter to every registered professional listed in the Yellow Pages of the phone book. Beware of these. The competition in such cases can be among several hundred professionals. Such an approach often indicates a seriously inexperienced client, and inexperienced clients have a habit of making selections on a basis of inappropriate criteria.

For example, a few years ago Communications Satelite Corporation undertook to engage A&E services for its new operations center outside Washington, D.C. The mechanics of screening professionals was handled by Comsat's Purchasing Department. From the outset the purchasing agents appeared to approach the matter very much as though they were buying systems proposals from the electronics industry. The job was widely publicized in order to encourage "competition." All interested firms were sent exhaustive questionnaires and were referred to as "bidders" in the correspondence. When some concerned professionals tried to contact Comsat's purchasing agents for clarification of the ground rules, they were told that no individual conversations would be permitted; all questions were required to be submitted in writing; and all answers would be circulated simultaneously to all "bidders." Eventually the message did get through and Comsat issued a number of "clarifications" which recognized the principles of professional selection. By then, however, a great number of firms had been put to tremendously wasteful effort to answer Comsat's questionnaire; other firms had given up in disgust; and Comsat itself was deluged with volumes of required but largely useless information. In the end, Comsat selected an eminently qualified, nationally known firm which it could easily have located by a much more simple selection process. This example is not used to embarrass Comsat or second-guess their selection. It simply illustrates the frustrations of pursuing a client who doesn't understand how to select professionals.

If the professional firm is really lucky, it may encounter a live prospect before it has competitors and while the field is relatively clear.

The difference in how contact is made with a live prospect is not really important, except in terms of mathematical odds which are discussed at the end of this chapter. The message of this chapter is what should be done when the quarry is in range.

Don't start the hard sell yet!

This is a time for strategy research. If you are going to sell yourself effectively later, now is the time to learn five things:

1. Who is the Prospective Client?

A working acquaintance with the prospect organization and the players involved is exceedingly helpful to good salesmanship. The quickest source of this is often an Annual Report—if one is published—and most public libraries, banks, or stock brokers will have them on file. Public records will tell almost anything one may want to know about government agencies. For educational institutions the "American Universities and Colleges" directory gives a brief history, the name of the president, and a list of the recent major buildings. A list of standard references worth having in an office or readily accessible is given in the Appendix. Even the busiest professional can send a secretary to the library to transcribe this data and bring it to the office.

In addition to knowing the client's business, it is important to know who the people are who run the organization. All sorts of annual reports, catalogues, and directories list these names. The primary value in knowing names, of course, is in the cross-referencing. For example, perhaps a director of XYZ, Incorporated, is also on the County Council for which the firm recently did a project. Or the president of the professional's bank is a trustee of the college where the firm is seeking work. "Namesmanship" is a fascinating game, and after one plays it awhile one learns invaluable information about the establishment in any community.

At the national level, *Poor's Register of Corporation Directors* cross-indexes all related affiliations of every corporate director.

Perhaps the easiest quick-reference source of background data and names on a whole range of probable prospects is Dun & Bradstreet Service. Each D&B Report gives a brief history of the organization and a complete list of all officers and directors with a brief biography of each. D&B Reports are copyrighted and the only legal way to see them is to subscribe. The basic rate is about $800.00 for 100 reports per year; and therefore only larger organizations will justify the service, but it is a great convenience to sales research for those who can afford it. If you don't subscribe and need only one or two reports a year, these can often be obtained from a friendly banker or broker who does have a subscription.

2. What is the Program?

Every client who is serious about hiring a professional has some sort of a program in mind. And wise professionals will know all they can about that program before they start selling any services. Some of the things the professional ought to know are:

How did the client decide what it wants to build?

Are the program requirements to be developed by the professionals who are about to be hired?

If a detailed facilities program has already been prepared, is this to be rigidly followed or will the new professionals be able to contribute to it?

Is there a budget? How was it arrived at, and how fixed is it?

How will the project be financed? Is construction money available, or will money have to be raised after preliminary design?

How will the project be built?

Obviously, the answers to these questions can have a good deal to do with how the client may select professionals. From the professional's standpoint, the answers can do much to help define the scope of services to be offered.

Never assume that the client knows what services are needed. The client may say what is wanted, but a good professional will never start there. Quite often preliminary services such as programming, site selection, master planning, and fiscal feasibility studies are called for before the client should begin to plan a structure. The professional who takes the trouble to research the client's problem before selling may often be able to open the client's eyes to possibilities and problems that have not been considered. And when that happens, score many points for the lucky professional.

3. What is the Site?

The professional should find out where the site is, and should visit it.

Most professionals agree that the character of the site has a very significant bearing on the design, quality, and cost of any project built on it. Yet it is surprising how many firms will pursue commissions without knowing anything about the site; and later, when difficulties arise, these firms will claim the job was severely limited by "site problems" or a "bad site."

On the other hand, the prospective client has often selected a site on the basis of very incomplete or naive criteria. While screening professionals the client may welcome comments—good or bad—about the site and the influence it may have on the project to be designed. At the very least, such comments show the prospect that the professional has taken the trouble to become fully familiar with the assignment. At best, the professional's comments about the site may do much to demonstrate a philosophy of approach, and thereby help convince the prospect of the professional's suitability for the assignment. In extreme cases, comments about an adverse site may convince a prospect to take a whole new approach to the assignment. The professional who gives such advice will invariably gain respect, and often will get the job.

The point to be emphasized in strategy research is the importance of identifying and inspecting the site. The time to present the prospect with comments about a site is discussed in Chapter 13.

If the prospect does not have a site, the professional will want to know whether the offer of services should include a site selection survey.

4. How Will Professionals Be Selected?

Strategy is all-important in effective selling, and the prospect controls all the ground rules that influence strategy. These include:

How many firms will be considered.

How they will be initially screened (by written submittals or interviews, or both).

Who will do the screening.

How many firms will be considered in the final interviews.

Who will do the interviewing.

What questions the interviewers will want to discuss.

Where the interviews will be held. (Is the room suitable for showing slides?)

How long the interview will last.

Will price proposals be requested prior to selection? If "yes," how much weight will be given to the lowest price?

How the final selection will be made.

Who will make the final selection.

No successful salesperson ever plans a campaign without knowing these ground rules.

Consider the complaint of the firm that walks into an interview carrying a carefully prepared portfolio of 8 X 10 pictures and finds a committee of 25 people waiting to see them. Or their embarrassment, the next time, when they go to an interview equipped with projector, slides, and an oversize screen, and then find one person waiting to talk with them around a desk in an office which cannot be darkened.

It is very common for interviewing boards to stipulate in advance how much time will be allotted for meeting with each firm. It is equally common for firms to prepare formal presentations which virtually fill that time, and then be thrown into unseemly confusion when the interviewers start the meeting by asking prepared questions.

At the other end of the scale are instances when firms make a fully effective effort to sell themselves, but to the wrong people. Quite often the written submittals are screened by a staff committee which has little or no voice in final selection. Interviews are frequently conducted by a committee which may only recommend to a higher authority, not decide.

It is entirely proper that the professional firm being considered for an assignment should know the mechanics of the client's selection process. No reasonable client is shy about revealing such information.

5. What Is the Timetable?

There are two timetables that are important:

a. The project timetable. Obviously, an extreme timetable will influence a professional firm's ability to handle the job.

b. The timetable for selection of professionals. This ties in directly with the other aspects of strategy discussed above. From the sales point of view it is particularly valuable to know the "consideration time" that may be allotted between receipt of written submittals and selection of those to be interviewed; and the time between the end of the interviews and the final decision.

Knowing these timetables may open up valuable opportunities for follow-up submittals, personal contact, and other effective techniques for salesmanship. When the timetable is not known, a great deal of effort can be wasted. No matter how well intended, a firm can make itself look perfectly ridiculous by continuing the hard sell after the decision has been made, but before it is announced.

It should now be apparent why all the information outlined in these five questions is worth knowing. Obviously, the answers in many cases reveal a great deal about the services the prospect is seeking and the type of firm that may be selected. And the answers can have a profound influence on the sales tactics used by anyone seeking the commission.

Now the obvious question: How does one learn all this, when the letter of inquiry from the prospect does not say?

Quite simply: Interview the prospect.

Contact the highest level person who can be reached in the prospect's organization, and ask. Don't worry if the person is removed from or below the level of those who will choose the professionals. The answers are what is important.

The approach is thus:

> We have your inquiry about the new South County project, and I would like to know more about your plans so that we can decide whether we are qualified to offer our services.
>
> May I ask you a few questions, or may I come and see you?

Nine times out of ten it will get you an appointment.

But remember, if the prospect is receptive to questions, don't try to sell anything at this stage. This is your opportunity to interview the prospect before you are interviewed. The answers may make the prospect a sitting duck for some effective sales pitches, but don't tip your hand. Save the salesmanship for answering the formal questionnaire or until the interview. Then you can surprise the whole committee with your perceptiveness of the prospect's problem.

STRATEGY JUDGMENTS

After getting the answers, there are three important questions the selling professionals must ask themselves:

> Do we want the job?
>
> Have we any chance of getting it?
>
> If we go after it, what is our best strategy to win it?

The answers to the first two questions deserve much more serious consideration than most design professionals have been accustomed to giving. A large majority of architects and engineers, when faced with any live opportunity, no matter how remote, are tempted to say: "Let's submit for it. After all, it's such a long shot we really don't have much to lose." In practice, there is a tremendous cost in losing. Internally the energy that is spent putting together long-shot proposals inflates the marketing budget and, more importantly, detracts from the kind of maximum energy that should be applied toward the ones you really can win. Externally, it is never wise to let a client reject you because you have no business trying to get the job. No firm's image is enhanced by indiscriminate scatter-shooting at every target in sight. Perhaps most important is the fundamental need for good judgment throughout if you expect to succeed at the business development process. If you don't have the judgment to walk away from the leads you can't win, how can you be sure to trust your judgment on the assignments you should win?

There is no faster way to gain respect for one's professional judgment than to withdraw from consideration for a job which is not your cup of tea. These letters are examples of very good salesmanship:

Dear Prospective Client:

We regret very much that we cannot accept your invitation to be considered as architect for the proposed new XYZ structure. Unfortunately, our organization is fully committed to other projects for the next six months, and I know if we were to be selected that we could not meet the timetable you have in mind.

We are most grateful for your inquiry, and hope when future projects arise you will give us an opportunity to be considered at that time.
Sincerely,

Or:

Dear Mr. Prospect:

We are highly flattered to have your invitation to be considered as engineers for the new River Bridge, an assignment which will certainly require professional services of the highest order. However, after considering all implications of this challenging assignment, we do not feel we should ask for further consideration.

We are a new and growing firm, and although we are most anxious to have the opportunity to provide engineering services for your Department, we must be realistic about our current capacity. To properly design the River Bridge would require a staff several times our present size, and we do not believe we should try to tackle such an assignment until we have grown a good deal larger.

In the meantime, we hope you will keep us in mind for projects of somewhat smaller scope.
Sincerely,

The moral is always to put the best foot forward. If it looks like one has no chance of getting a particular job, don't go ahead and strike out. Every time a board of peers considers and rejects a professional, your batting average declines. And when the average gets too low, no one will give you a chance.

On the other hand, if the answers to the questions discussed in this chapter show the job to be just the sort of opportunity you want, now it is time to chart your strategy.

The strategy development process can be diagrammed by the following matrix.

Strategy Research	leads to	Strategy Judgments	=	Presentation Approach
i.e., What can we know about the assignment, the client, and the competition?		e.g., What do we decide about what we know?		What and how will we sell?

The decisions called for in the middle of this equation are crucial to the success of every selling effort. It can be assumed that all the competitors will have access to the same strategy research data. But it is highly unlikely that any two competitors will interpret all this data the same way. The judgments you make at this stage give you the greatest opportunity to establish your uniqueness in front of the client and thereby walk away with the job.

It is good practice at this stage of planning a pursuit to review the following list of questions.

Strategy Judgment Checklist

1. What is the real job?

2. What do we know about the prospect (personality, position, decisionmakers, influencers, preferences, etc.)?

3. What do we know about this prospect's project:
   —The objectives
   —Their view of the problem
   —What they need from us (scope of services)?

4. What is our view of:
   —The prospect's objectives
   —The prospect's problem
   —What the prospect needs
   —Alternative approaches to the project?

5. What does the prospect *expect* to hear from us?

6. What does the project *need* to hear from us?

7. What project team will be involved?

8. How will we present our approach?

9. What is the single most important message we want the client to know about us?

It should be clear that none of these questions lend themselves to clear, absolute answers. Each provides only an opportunity for a choice. This is why they are called strategy judgments. You have the chance to adopt virtually any sales strategy you choose. The correctness of your choice in terms of whether you win or lose will depend, more than anything else, on the quality of your original strategy research.

If you know everything there is to know about the prospective client, the assignment, the competition, and the process of selection, then it is time to start selling!

from: **the Coxe Letter**

CASE STUDY #80: ASKING HARD QUESTIONS GAINS COMPETITIVE EDGE

Cannon Design Inc. is a 170-person diversified A / E practice in Grand Island, New York, with a solid track record in hospital work. A little prospecting by an associate last spring resulted in a recent invitation to be interviewed for a New Jersey hospital expansion study.

The firm's Vice President of Marketing, Mr. Clare Ross, says: "We were competing with two top firms and our first concern was whether we were just in there as window dressing. The best way to find out is to ask." Ross visited the administrator in charge of selection and learned specifically that he was impressed by Cannon's hospital renovation experience. Ross also asked for the names of the decisionmakers, their occupations, and their business addresses. Then, with a new Regional Manager, Don Brennan, who Ross was breaking in, they cold-called each member of the selection committee. "Everyone saw us," notes Ross, "and seemed to be surprised that we called on them. No other firm had been to see them."

"We wanted to identify the hospital's needs as each perceived them, and we wanted to know what our competitive advantages—or disadvantages—were. We asked each of them, 'How important is it for your architect to be local?' (the competing firms were based in New York and Philadelphia, close enough to be considered local); and 'Why do you want to consider Cannon Design?' Proximity wasn't an issue for anyone except for an architect on the Committee. We found out we were there because we were qualified."

Ross continues, "Then we asked what problems they saw, and how they felt about the project. The problems were pretty consistently perceived. They had a master plan that wasn't responsive to their needs, and they recognized that they needed professional help to explore what their options were. We also found out that money was important—they needed a plan for phasing expansion over time."

Prior to the interview, the firm constructed a schematic model of the existing buildings and alternative new segments, using solid wood blocks. Slides were then taken of the model to demonstrate phased expansion possibilities.

Five people from Cannon went to the interview, including Brennan. ("The marketing person has got to be there," says Ross. "He's the only face everyone will recognize.") Slides of the model were shown. "The committee immediately understood what we were saying," comments Ross. "We showed how they could satisfy their total needs over several years, as they could afford it." Cannon got the job and expects to be the hospital's architect for some time to come.

Ross concludes, "The people who are best able to define and respond to the client's needs are going to get the job. The way to find out is to ask. When you've got information on biases, concerns, and interests, you have an advantage."

CASE STUDY #101: EARLY HOMEWORK GETS THE JOB

Whiteside, Moeckel and Carbonell (WMC), a 22-man Wilmington, Del. firm, first got wind of plans to upgrade Amtrak's Northeast Rail Corridor in 1975, and soon learned that the program would include a substantial renovation of the Wilmington station. The decision was made to pursue the project, and Henry E. White, a senior associate, undertook to find out more.

Fortuitously, one of WMC's contacts from another project was part of the team that landed the management contract for the Northeast Corridor Improvement Project. Although not in a policy-making position, he was able to speculate on some of the important factors to be weighed in awarding commissions for the work. Thus, "We guessed what was happening before it happened," says White.

One of the priorities of the program was minority involvement. White and colleagues made numerous trips to Washington to get to know the appropriate officials and to learn the requirements to satisfy minority employment standards. The first fact they unearthed was that WMC itself qualified as a minority firm, since partner and 50 percent owner Joseph E. Carbonell, Jr. is of Hispanic heritage. White then requested the names of minority firms qualified to be retained as consultants. He contacted about 35 of these firms, 15 of which responded with 254s.

The minority goal of 25 percent of the total project fee was fulfilled by the inclusion of Ysrael A. Seinuk for structural/bridge engineering services. The entire consulting team was assembled by White three weeks before the project's general announcement in Commerce and Business Daily—and well before some competitors even got started. It also included Henry Adams Inc. (mechanical/electrical engineering); Giorgio Cavagliere (historic preservation); Van Demark & Lynch Inc. (civil engineering); Wilbur Smith Associates (trackage/traffic); Construction Environment Inc. (geotechnical engineering); Cerami Associates Inc. (acoustics); Donald R. Knox Inc. (landscaping); John O. Meadows Associates (cost control).

On a subsequent visit to Washington, White learned coincidentally that, in addition to a minority employment official, two members of the selection committee were already acquainted with some of the firm's work.

"Washington," muses White in recalling the popular epithet, "really is the biggest small town in the world."

WMC was included on the five-firm shortlist after project officials had eliminated the remainder of the 70-odd responses. The interview format was to include precisely 30 minutes of presentation, followed by a 15-minute question-and-answer session. It was decided that the best strategy was to emphasize WMC's single-point responsibility; and to concentrate on a cogent plan for solving the problems at hand, rather than to recapitulate past triumphs of design.

Of particular import were traffic and circulation patterns in and around the station. To illustrate the team's grasp of the situation, WMC hired a helicopter for a day to obtain aerial photos illustrating specific problems to be considered.

White also felt it essential to provide a leave-behind brochure, especially since WMC was scheduled to be the first of the five teams to make its case. The brochure included a summary of minority involvement, the qualifications of team members, and even a schedule of the travel times required for consultation.

After an all-night rehearsal in Washington to hone the presentation, the team was ready.

"We began by explaining that we would not, at great personal sacrifice, display our favorite 400-or-so slides," says White. "It broke the ice and I explained we knew the committee thought we were capable, or we wouldn't be there.

Carbonell briefly introduced the team members, after which project-manager-to-be White took charge. Included were two or three slides and some concise remarks from three of the consultants, followed by some of the spectacular aerial photos that "knocked the committee members out of their seats," according to White.

White concluded the presentation with special dramatic emphasis on the fact that it had lasted exactly 30 minutes—almost to the second.

"You could see the amazement register on their faces as they checked their watches," recalls White. "It must have been on their minds as they observed the four remaining presentations."

One month later, the commission for the $15 million project was awarded to WMC (soon to be renamed Moeckel, Carbonell and Partners) and its team.

11

COURTING

*"43% of salesmen make one call and quit;
25% make two calls and quit;
12% quit after three calls;
80% of all sales are made after the fifth call!"*
                                    Sales Management magazine

The elements of the business development process described up to this point are *preselling* activities. Choosing markets, building lists of potential contacts, bird-dogging to find prospects, and strategizing the leads that are turned up are all phases of the process that sets the stage for pursuing a specific commission.

Direct selling begins when one has a target in sight and has developed a strategy for pursuit. The formal activities of this selling process—especially the paperwork submittals, presentations, and closing techniques—are described in the remaining chapters of this section. All along the way, every selling pursuit involves courting the prospective client in order to develop rapport and confidence.

In practice, some firms initiate a high proportion or all of their marketing at the courting phase. These firms know of certain active clients with whom they want to work, and they focus their business development effort on getting known by these clients.

Wherever it occurs, the process of courting a potential client is a distinct and manageable business development activity. It must be made clear that this type of courting is not analogous to the "who you know" kind of marketing that was so prevalent in the design professions years ago. There will always be some work awarded on personal favoritism—that is the nature of human interaction—but the design marketplace has changed drastically in the last two generations, and today's clients demonstrate time and again a high degree of openness to new approaches and new faces. "Who they know" matters not in the sense of who will get the job, but "who they don't know" may have a lot to do with who will *not* get it.

Courting, by definition, is the process of establishing a level of professional relationship that makes the client comfortable with the idea of working with you—usually well before a decision is made on who will get the job. In every professional service, people must be comfortable with whom they are working if either party is going to get the most out of the professional relationship. Thus, every effective business development program must include an understanding of how and when to build those relationships and who must do it.

BASIC COURTING TECHNIQUES

There are essentially three requirements to successfully court a professional relationship:

Be known.

Be credible.

Be useful.

The routes to becoming known are simpler than many professionals make them. One way, of course, is to be introduced and this can be effective if the right person makes the introduction. Since credibility is also a key part of the courting process (see below) it is important that the person making the introduction have equal or greater credibility than that by which you wish to be known. Otherwise, you may have an uphill struggle to establish a strong posture with the prospect you are courting.

Lacking an introduction, the easiest way to become known is to introduce yourself and then reappear regularly and often enough so that your name—and even your face—are remembered and recognized. Federal agency contracting officers, for example, report they see hundreds of one-visit design professionals, who rush in when they are hungry and show their disappointment when they find there is no immediate work to be had. It is not surprising, therefore, that clients pay little attention to those they meet on a first visit. Only after two or three visits will they begin to take you seriously and start to offer genuinely helpful advice about what is going on in the organization and what is coming up.

Getting known is just this sort of process with all prospective clients: You must reappear at discretely regular intervals—generally 60 to 90 days apart—and be patient enough to keep at it for six to nine months until you are readily recognized.

Becoming credible is the next objective, and this involves both what is conveyed during the get-acquainted visit, and who does the conveying. If the courting is done by a principal professional who is prepared to be the engineer or architect for the client, the credibility conveyed will be quite different from what it might be if the caller is seen only as a sales representative or bird-dog for the firm. In practice this match of courter and prospect is a critical consideration and is the reason why there is so much misunderstanding in design firms about how much of the marketing effort can be delegated to nonprofessional bird-dogs. The role of the bird-dog is to get information and find the leads. This does not mean that the bird-dog will be a credible courter thereafter. In some markets, especially public agencies, bird-dogs can build reasonably effective relationships with clients; but in other markets, especially in the private sector, only card-carrying registered professionals are given serious credibility. In any market, if the design firm's representative doing the courting is seen as "only" a salesperson seeking work, credibility will be hard to establish. Whoever is doing the courting must make genuine efforts to project a useful benefit to the prospective client, if any sort of credible professional relationship is to be developed.

There are many ways to project the kind of benefits that will gain credibility. One way, of course, is to be an expert in the project type and make it clear to clients you know their business and may have something they can use. This is why a great deal of credibility is given to architects or engineers who appear on programs at client trade associations, write articles for client trade magazines, or otherwise become recognized in a client's field. When these professionals go courting, they are always way ahead of the generalists in establishing credibility.

The generalist, however, is not powerless. If the firm has particular strengths in personnel, location, process, cost controls, or whatever, these can be conveyed during the courting process to gain credibility. A good practice when courting is to look for something specific to take to the prospect on each visit—perhaps a printed project report that might be of interest, a reprint of a trade magazine article about the firm's work, or a new edition or supplement to the firm's brochure. If these discussion elements are carefully chosen to build a case that interests the prospective client, you will, over time, gain considerable credibility.

The last, but by no means the least, route to credibility is through civic or professional associations. If the client sees you perform well at something—even a wholly different activity—you will gain stature in the client's eyes. This is why professionals who are active on civic boards or in professional society affairs do so well when they seek work from clients who are aware of these roles. It is not the contacts that are made in public / professional service that are so valuable; rather it is credibility you gain as one who has demonstrated the ability to lead and get things done.

The final element of the courting process is being helpful. If one has become acquainted with a potential client, and established credibility, there will come inevitably occasions when you can help that person do their job. It might be going to look at a site and offering some gratuitous suggestions; it might be helping to write a request for proposals; or it might involve something personal such as helping the person's son get a job interview with a contractor. When courting reaches the stage where the prospect is comfortable asking for advice or a favor, and the professional is comfortable giving or granting it, a mature relationship has been established that will inevitably yield mutual benefits over the long term. That is what courting is all about.

SHORT-TERM COURTING

There are, of course, frequent occasions in the selling process when there is need for shorter-term courting of members of a board or selection team who are considering professionals for a particular job. The process of contacting these people do develop relationships prior to a vote on selection is no different than the long term courting described above. Only the time frame is much shorter.

The first goal is to get known. The difference in a short term situation is that you are likely to be remembered after even a single visit because the client has a current reason to want to know you. Nevertheless, more than one visit may be very helpful. The key point in these visits is to project credibility and benefit, not just hunger. If, for example, you are calling on members of a hospital board that is considering architects, it is important that the agenda for the call be to establish credibility, not just to ask for favorable consideration. The best calls in this situation are those where you can be useful at the same time. If, through strategy research, you learn that the members of the board have a particular concern or interest and your visit can focus on giving them some information they seek, you will win points for credibility and usefulness at the same time. The story is told of a marketer who learned that a civic selection board was inexperienced and was preoccupied with being fair in its consideration of professionals. The marketer borrowed an interview score sheet from a federal agency, and took it to each member of the board with a comment like: "We know you face a tough selection because you are considering several good firms in addition to ourselves. I thought you might like to know how government selection boards do this, and I've gotten permission for you to use this form if you like." The marketer not only won points for being useful, but also thereby helped shape the selection criteria. The firm's interview was stacked accordingly. Meanwhile, the competition didn't find this out until it was too late, and their interviews all fell short.

PLAYING POLITICS

In some markets, especially in the public sector, there has long been a mystique about the courting process because of implications or suspicions that "politics" are also involved. Thanks largely to a growing public distaste for this way of doing business, and thanks also, ironically, to former U.S. Vice President Spiro Agnew who was convicted of taking kickbacks from engineers, the political factor in architect and engineer selections is considerably reduced from what it was a generation ago. The author's observation is that "politics" in all its unattractive or illegal ramifications is a factor in only a minor fraction of the market for professional design services today. This does not mean, however, that it is not important to learn as early as possible in the courting process whether or not you are dealing with an open professional process. As mentioned in the previous chapter, if the rules of selection require kickbacks to a politcal party, or off-color entertainment of the corporate vice president, no amount of straight professional courting is likely to be successful. In strategy research it is essential to ask as much as you can about whether selections are open or wired, and at the strategy judgment stage you must decide whether or not you are going to play the apparent game.

The vast majority of design professionals (and public clients) have washed their hands of playing "dirty" politics since the Watergate / Agnew-related scandals of the 1970s. But there is a volume of work where "clean" politics is still a factor, and this deserves consideration in planning a courting campaign.

In the public sector, "clean" politics arises when an elected politician is in a position to do favors for friends, and uses influence on behalf of professionals who may have worked for years on the politician's campaign staff. This is not in any way illegal, but it is politics, and if you wish to court political favor in jurisdictions where the game is played this way, you must make your friends and lend your useful support years before you may expect any return. On the other—and best—hand, not all this political involvement is selfish—many public minded architects and engineers seek credible positions in the public sector as a route to influencing legislation and public policy in favor of professional principles in which they believe. In these cases many end up serving in positions which exclude them from accepting public commissions. The point here is merely to emphasize that courting political position can also be part of a marketing strategy, and if it is done ethically and professionally it can lead to long term rewards.

In the private sector the practice of playing unclean politics is diminishing even more rapidly than in the public sector, but there is still a lot of work where commissions are awarded as favors to friends. The same rules of courting apply here as elsewhere: A friend can do you a favor only by being credible and useful, otherwise the reference isn't likely to carry any weight. If you do ask a friend to help you win a job, the obligation to execute it well is the greatest of all because you must live up to the credibility of both the client and the friend who went to bat for you.

In essence, the political courting message today is that "who you know" is only useful if you have established substantial credibility in the first place. And if you have the right kind of credibility, you may not need political help to win most of your work.

THE COMMAND GENERATION

There is in every community and every business a prevailing generation level at which leaders begin to assume the top positions in their organizations. For the most part this "command generation" is the group between about 45 and 55 who comprise the corporate, banking, and institutional presidents; leading developers, lawyers, and doctors; most active politicians; etc. It is the group which have "arrived" and have most of the reins of power, civically, economically, and politically in their hands. (There are exceptions to the age norms, to be sure, but they are often of those of whom it is later said they "peaked too soon" or "stayed too long.")

For architects and engineers, there is no substitute for being "in" with the command generation of their market. This is why so many firms flower when the principals reach this age.

In marketing planning, it is well to remember how these principals originally got "in" with their command peers. In most cases it wasn't accomplished just by walking into a new community or power structure at 45 and knocking on the door. Rather, the position in the command structure was earned by apprenticeship in civic / community / business activities at an earlier age. Old school ties may have helped here and there but the best peer acceptance was earned when the future leaders were all in their 30s and worked together on United Fund boards, political campaigns, and/or played ball together after work at the Y. At this age, when everyone was struggling up the ladder, they all got a chance to compare leadership talents. Those who passed the test of peer acceptance at this age earned charter membership in the next command generation.

This explains why it is often so difficult for the founder/marketer in a design firm at age 60+ to pass the marketing torch to a 45/50-year-old associate whose role until then has been inside the firm making it work. The time to begin positioning future marketers is when they are in their early 30s, by giving them the encouragement, time, and expense accounts to become civically/professionally active among their peers.

To make all this happen on schedule an established firm must have the long-range vision necessary to push into the forefront talented young professionals who haven't even decided yet whether to stay in the firm or branch out and become competitors. A new young firm, on the other hand, must have the long-range patience to "serve its time" awaiting the day when it will grow up with its command generation.

Either way it's all part of the courting equation. This chapter has not been written to imply that the only way to market professional services is through courted relationships. Far from it. The great majority of work goes first to the firm the client believes is best qualified for the job. But when it comes to a new firm being given its first chance, or when an established firm gets down to the short-short list against an equally qualified competitor, the degree of professional relationship that exists between the architect or engineer and the client may have a lot to do with the final choice.

If that is understood, now is the time to talk about how to sell a firm's qualifications.

from: **the Coxe Letter**

CASE STUDY #1: GETTING PROSPECTS TO VISIT THE OFFICE

A medium-size architect/engineer in Central Pennsylvania has developed a novel method of using their off-the-beaten-track offices as a major tool in developing new business.

The firm solicits school boards and town and county councils over a wide area, and uses a Winnebago motor home to transport prospects to and from visits to the firm's offices.

A typical tour will begin about 4 PM when an entire school board is picked up by the Winnebago. It is specially fitted for lounging, and cocktails are served from the on-board bar during the drive. (They find that $2\frac{1}{2}$ hours or 100 miles is the practical maximum.)

On arrival at the offices the visitors are walked around to "see the scope of the firm and how we operate." Key department heads are on hand to explain in detail how their sections operate. Then the group gathers in the main conference room (decorated with appropriate models) for questions and answers. The office visit is timed to last a maximum of $1\frac{1}{4}$ hours unless the visitors specifically want more time.

The Winnebago then takes the group to a local restaurant for a first-class dinner and an opportunity to mingle with selected principals of the firm. The return drive is always scheduled so the visitors are returned to their homes before midnight.

The firm leases the Winnebago on an as-needed basis from a local dealer, and considers the cost minimal in comparison to the value.

A firm spokesman attributes several commissions to the technique including particularly a school 85 miles away where the District had specifically asked architects not to make personal calls on board members, but had no hesitation in accepting the invitation for the office visit. "We feel the most important thing we have to sell is our company," he says, "and in selling the company, people just have to see it."

CASE STUDY #91: CLIENTS ARE FOR KEEPING

This year Gresham & Smith, a 95-person A/E firm in Nashville, started a "Client Appreciation Program." After analyzing the sources of new business last January, the partners recognized that most of the work was coming from repeat clients and referrals—and that these people deserved more attention than they were getting.

As a result, Director of Marketing Jerry Guy decided that the "most important" people should be identified and contacted four times a year in a systematic program. He put together a list of some 100 names, including clients and "people who help us find clients" such as developers, school boards, metropolitan government people, consultants, and bankers.

Guy then circulated the list to the 24-member business development group for additions, and asked the core marketing group (5 partners, 2 associate partners, and 3 professional associates) to indicate whom they'd like to take responsibility for contacting. Each of the ten is responsible for 5 to 20 individuals. At four contacts per year, each makes 2 to 8 client contacts per month.

Orlean Gibson, Marketing/Communications Coordinator, developed a system to implement the program. Once a month she circulates a reminder memo which reads, "You are scheduled to make contact with the following clients this month" and lists the names, companies, and phone numbers. The memo also lists the kinds of contact for their choice—visiting their office, inviting them to visit Gresham & Smith, a lunch out or at the G&S office, a social or recreational outing, sending tickets to a cultural event, sending a reprint, writing a brief personal note, a phone call, sending a pertinent newspaper or magazine clipping.

After making the contact, each fills out a contact record (in triplicate) documenting who was contacted, when, and in what way. This is returned to Gibson, who puts one copy in the tickler file for the next round, one in the client folder, and one in the contacter's folder.

Guy reports that about three-fourths are made within the month. Those who fail to meet the deadline are reminded by partner Batey Gresham, and within the next two weeks virtually everyone has done their work.

"Client reaction has been very favorable," notes Guy. Already one contact seems to have been instrumental in getting work: "Another fellow and I were playing tennis with a developer," he recalls, "and after the match he told us he had a new shopping center he wanted us to design. I don't think we would have gotten it if it weren't for the client contact program. He said to us, 'You folks have shown interest in doing business with us, and we really appreciate it.'"

CASE STUDY #105: SEMINARS BUILD RELATIONSHIPS; RELATIONSHIPS LEAD TO JOBS

When the Los Angeles office of Syska & Hennessy, Inc., Engineers, started a series of seminars a year ago, there was concern about attracting attendance. Now, says William A. Feathers, Director of Development, people call to be invited.

The series was conceived "to broaden our base of contacts, and to allow our guests to broaden theirs," says Feathers. "Engineers traditionally market to architects, but of course that's only one of the relationships which must be solidified before a project can be executed." Accordingly, Feathers developed an invitation list including real estate brokers, land owners, developers, financiers and design professionals of various disciplines. The list presently numbers 300.

The seminars are scheduled about every six weeks. Designed to be an "on-the-way-home" event, seminars begin with a cocktail reception in Syska & Hennessy's board room at 5 P.M. Speakers—which have included economists, space planners, investment analysts, and developers—address the guests briefly, followed by questions and answers.

Syska & Hennessy's most recent speaker was Henry Lambert, president of Continental Cities Company, Inc., the real estate development arm of Reliance Group, insurance and land title companies. Roughly 70 persons have attended each of the seminars held to date. The opportunity to meet the speakers as well as each other, Feathers says, is the chief cause for the program's success.

The program has generated considerable good will, says Feathers. More tangibly, a recent seminar served as the catalyst for Syska & Hennessy's inclusion on The Luckman Partnership's design team for a Dillingham Corp. development in Long Beach, CA. The project will consist of two office towers, each with approximately 250,000 sq. ft. of gross leasable area.

12

COMMUNICATING ON PAPER —LETTERS, QUESTIONNAIRES, AND PROPOSALS

"Keep the paperwork to a minimum. If you don't have an overriding need to say it—don't."
Stephen A. Kliment AIA

After contact is made and a project of genuine interest is identified, the next step is to let the prospect know if the firm is interested in being considered for the commission.

How well this is done—normally by letter and/or by completing a questionnaire—is critically important.

At this stage the prospect is likely to be screening some 15 to 100 professionals in order to choose three to six to be interviewed. Thus, the competition is now at its greatest, and the prospect's judgment is most objective. Every professional in the running is going to do their utmost to survive this screening and get tapped for the short list.

Although the variations in technique are limitless, there are really two ways to survive this screening: One is to use an "inside track"; the other is to rely on the merit of the written submission.

This chapter deals with how to do it on merit. However, a preliminary word about the "inside track" may be appropriate. Use of the inside track to gain consideration for professional services is a time-honored device, and for those who can find such a route it may well be the best approach to gain consideration, but it can also be a trap.

Ask anyone who has served on a few selection committees how the candidates for interview were selected, and often the explanation will go something like this: Three or four were selected because their paperwork submittals indicated substantial prior experience in the building type; one or two were special cases selected because of fame, reputation, or a particularly unique approach that the committee was curious to see; and lastly there were one or more "courtesy" interviews to firms having "inside tracks" because they were friends of someone in the organization or because they had done prior work for the client, or etc. In such cases, it is rare that those relying solely on the inside track will survive the interview and win the commission. If one truly has an inside track one should be able to use it to persuade the client to make a direct selection, without competition. Otherwise, unless the whole selection process is a sham (and unfortunately this is still sometimes the case), the inside track ends at the door of the interview. And in many cases the inside track can work against an otherwise qualified firm. Those on a selection committee who feel they have a potential bias in favor of one firm often bend over so far backwards to avoid appearing unfair that they let their good judgment be overridden by an uninformed minority.

In a conventional, merit selection process the purpose of the paperwork phase is to get closer to the job, e.g., to narrow the choice but not to win it. The degree of importance of this phase has changed considerably since the first edition of this book was written. At that time the great majority of selections followed this sequence:

$$\text{Long List} \overset{\text{Qualifications}}{\underset{\text{Paperwork}}{>}} \overset{\text{Short}}{\underset{\text{List}}{>}} \text{Interviews} > \text{Selection} = \text{Negotiation}$$

With the introduction of price competition into the selection of design professionals, more and more clients have started calling for specific proposals which include both a defined statement of services and a commitment to fee. In a literal sense a proposal is a completely different document from a qualifications presentation, but frequently they are combined. At this writing, a significant number of misguided clients are taking full scale proposals from a long list, putting everyone—including the client—to an unreasonable amount of paperwork.

The better process, used by more informed clients, involves a sequence more like this:

Long > Qualifications > Short > RFP > Proposals > Interviews Selection
List   Paperwork   List

Only time and experience will tell how much paperwork will be required eventually in the professional business development process, but it is clear the amount is increasing. As this continues to happen, two cardinal truths for the handling of business development paperwork become more and more apparent:

1. Boilerplate is less and less effective as competition increases. The way to stand out in a crowd is to dress specifically for that crowd— e. g., write presentations custom tailored to each selling situation.

2. The only effective way to know how to custom tailor paperwork is to do strategy research (Chapter 10) before the writing begins.

Thus, the preparation of effective paperwork presentations is taking more time and more care in professional organizations than ever before.

Design firms must be prepared to market themselves both through qualifications statements and through formal proposals. This chapter will endeavor to treat both.

THE QUALIFICATIONS PAPERWORK PROCESS

As a hypothetical, but not uncommon, example of the conventional process, consider the problems which face the professional selection committee of a local municipality. As a public agency it will feel morally obligated to invite submissions from all interested firms—and perhaps a dozen or more will have filed letters of interest during their bird-dogging activities. In addition, the committee may feel it politic to invite one or two firms whose principals live in the district, and also the firm which did the last project for the district five years ago. Finally, if the committee has particularly strong feelings about the type of facility it wants to build, it may invite one or two nationally known firms who have previously designed such projects. All this adds up to, perhaps, two dozen candidates.

Then the Committee agrees to devote two evenings to interviews. If candidates are offered one hour each, this schedule would normally allow time for no more than six interviews.

How will the six be selected?

In most cases, the selection committee will request written submittals of brochures and questionnaires, and these will be reviewed to shorten the list. Thus, the quality of the paperwork submitted becomes very important.

At the outset it must be emphasized that the real purpose of the documents submitted at this stage of the business development process is to get an interview, not to get the job. Now is the time to communicate qualifications, not specific services. The difference is perhaps easiest to understand if one has ever served on a professional selection committee. (Every professional should try to have this experience at least once, by volunteering to help a civic board, church, or similar group. Of course, in such cases, the professional will normally be disqualified from seeking the commission under consideration, but the experience will repay this sacrifice tenfold.)

Picture a committee of six people around a table, passing among themselves several dozen submissions. On the first round, each member will divide the group into two piles: those to be considered further, and those to be discarded. At the end of the first round, probably one half to two thirds of the firms will still be in the running. The committee will then discuss its reasoning, and will probably agree on some common ground rules for a second round. The process continues for as many rounds as are necessary until the list has been narrowed down to the desired number.

After one has read the answers of 25 firms to the same questionnaire it becomes obvious that the only way to narrow down such a group is on the basis of very specific criteria. The larger the screening committee, the more specific the criteria are likely to become, for each member will develop his or her own ideas. The eventual compromises will tend to narrow the criteria to those common among all members.

Thus it is essential that written submittals be explicitly designed to survive the screening. Presentations must not only sell the qualifications of the firm, but should also recognize the human chemistry of those who will do the screening. The dynamics of committee action usually tend to produce compromises, whereas an individual patron may stick to strong convictions. Hence, taking an off-beat approach to a questionnaire might work well with one type of prospect, but could be damaging with another.

Several generalizations can be made about the screening process:

Firms with previous specific experience in the building-type under consideration always rise to the top first.

If all firms give the same answer to a question, no matter *how* correct, the question becomes meaningless.

A stock brochure, by itself, will not receive the same degree of attention as individually prepared materials.

Graphic *elaborateness* of the submission is seldom an important criterion, and can induce negative reactions. Graphic *quality,* however, says much about the firm, and can be very important.

Brevity, with completeness, is an appreciated virtue.

The genuineness of the firm's interest in the assignment under consideration is very important.

The covering letter is possibly the most important document of all.

Most of these generalizations are derived simply from human nature. The psychology of persons faced with making a decision could occupy a volume by itself, and there are no absolute rules, but simple intuitive judgment can guide the selling professional.

For example, if the selection committee is composed of lay citizens, they will be interested in basic knowledge that might bore a committee of professionals. Thus, the written presentation to a school board will be quite different from the written presentation to the engineering section of the highway department.

In all cases, it is worth remembering that few committee members like to take sole responsibility for a decision. Committees tend to make "safe" decisions which, if challenged later, can be rationalized by putting part of the blame on external factors. Thus, any firm with prior experience in the project type under consideration will automatically rank ahead of one without experience. If the selection proves a mistake, the committee can always say that they were persuaded because the firm had done a successful project for someone else. The firm without prior experience must therefore make a special effort in its presentation to overcome this handicap.

In any presentation, it is critically important that some sort of individual personality come through the written material. Remember, it is a rare firm that is without equal in basic professional qualifications. If a selection committee has done its job well it is entirely possible that, at some point, it will be faced with choosing among several firms of almost equal qualifications. At times like this the nod may well go to the firm which says most persuasively why it wants the job.

The validity of these generalizations was very well illustrated a few years ago when an agency of the federal government set about selecting professionals for a $70 million administrative office building in Washington, D.C. The agency involved was not subject to the political criteria that often sway decisions on very large federal jobs, so the resulting process was a fairly typical example of the selection process.

The agency appointed a task force of its staff to develop selection procedures, and they issued a public call for submissions.

All firms who expressed interest were sent a two-page form letter which briefly outlined the program, described the site and some special conditions involved, and asked for presentations from professionals capable of providing all necessary architectural and engineering services. The standard government architect / engineer questionnaire was also requested.

The client received more than 125 submissions, and because many had interpreted the wording of the government's inquiry as a call for joint ventures, the submissions included material on architectural firms; structural, mechanical, and civil engineers; soils experts; traffic planners; food service consultants; and other special disciplines—more that 350 firms in all. Some of the joint venture names read like telephone directories.

The task force then set about screening this list down to five firms who would be invited to interviews with the senior officers of the agency. Composed of nine highly competent administrative and professional executives of the agency, it took over special quarters and devoted three weeks to scrutinizing the submittals.

It was no small job. The typical submittal was a joint venture of three firms. Their package included a lengthy joint statement of interest and qualifications, usually printed and spiral-bound together with the government forms for each firm—a minimum of 40 to 60 pages. Attached to this were the individual brochures of each of the firms, making a total pile four or five inches thick.

A number of firms were concerned by the graphic ungainliness of this, and devised elaborate bindings or burlap covered boxes to contain the various elements. And having gone this far with special constructions, these firms usually submitted the whole thing in triplicate.

Those firms who became concerned about the uniformity of letter-size materials went up in page area—all the way to 20 by 30 inches.

Several of the submissions were of such scope they were literally shipped to the task force by freight.

The task force endeavored to maintain the utmost fairness, and each member studied and read every last one of the submittals. To assist in the screening, the group developed a check list and each member gave numerical values to certain criteria as he reviewed the material. The aggregate scores were then used in narrowing down the list.

The task force apparently did its job well, for in the end the five firms selected for interviews were all established, nationally-known organizations with demonstrable experience in large-scale office buildings.

Of the five finalists, however, one had sent as its complete submission a one-and-one-half page letter. The letter was a superbly succinct, fully professional statement of the architect's experience in the building type, the size of the firm in relation to the scope of the commission, and expressed genuine interest in the job.

In the mass of other submittals, this letter could not help but stand out. So impressed was the task force by it, that they went to the trouble to telephone the firm to ask for the required government A / E questionnaire which, by the ground rules, had to be on file prior to the interviews.

The lesson is clear: If you say the right things in the right, succinct way, no committee will overlook your submittal.

It is perhaps anti-climatic to add that the professional whose superb simplicity won that spot on the short list did not, eventually, get the job, but this only serves to emphasize the point made earlier about the vital difference in approach between the *written submittal stage* and the *interview stage* in the business development process. The goal of qualifications paperwork is to get on the short list. Once one is invited to talk, a whole new set of selling rules apply, as is discussed in the next chapter.

It is one thing to say that submittals should be well written sales documents, but defining what this means is a very subjective process.

It would be impossible to anticipate the most effective answer to every conceivable question that might be raised by a prospective client, but there is a common pattern of information that must be communicated. Where questionnaires are involved, the most often used (or copied) are the U.S. Government Standard Forms 254 and 255 "Architect-Engineer Questionnaire." It is worth discussing these in some detail, not only because the forms are used by so many agencies but also because many of the principles apply to similar questions whenever they are asked by prospective clients.

U.S. Government Forms 254 and 255

All federal government agencies, and many state and local governments as well as some private clients, base their qualifications and selection files of architects and engineers on the General Services Administration's Standard Forms 254 and 255. These forms were adopted after several years of collaborative effort between the government and the design professions aimed at finding a fair way to compare firms on paper. The forms have two distinct functions:

Form 254 is a 4-page compilation of standard, nonperishable data that can be used to express interest in a particular agency. It is intended that firms file the form annually with agencies in which they are interested. The agency, in turn, can computerize or file the data by "experience codes" and thereby pull up a list of qualified firms whenever a particular project type is under consideration.

Form 255 is a longer, more flexible questionnaire that seeks information relevant to a specific commission. It is intended to be prepared and submitted separately in response to each project where an agency is seeking architectural or engineering services.

The Form 254 and 255 questionnaires (see Appendix) are so widely used by government agencies, and so much of their content seems pro forma, that some professional firms tend to fill them out largely with boilerplate and send them in with applications for all kinds of federal work. This is exactly the wrong approach.

One firm followed this mass-production practice in all innocence until one day a contracting officer for the Veterans Administration threw the form back across the desk and commented: "You will never get a hospital job from us with that form."

This comment was a particular jolt since the firm did 30–40% of its work in hospitals and thought itself exceptionally well qualified.

The VA officer then explained that the answers to Question 11 on the Form 254 ("Project Examples, last 5 years") listed the firm's projects chronologically and did not concentrate on hospitals. While this was an entirely accurate answer to the questionnaire, the particular VA contracting officer had long since decided to give preference to those firms which specifically indicated a concentration on hospitals. By revising the format of its Form 254 the firm was able to show just the experience the bureaucrat wanted.

The VA does not necessarily reflect government-wide policy, and that is just the point being made here. Another agency may take a completely opposite position. What this example is intended to emphasize is:

1. The Form 254 should be individually filled out for each different agency being courted.

2. The Form 255 should be very individually filled out of each different project being sought.

3. Once filled out, try to deliver the form(s) in person to the contracting officer concerned and ask if it has been completed in the manner the agency prefers.

In order to give the forms the special, custom orientation that agencies prefer, there are three questions on Form 254 and six on Form 255 that deserve special attention. These are:

Form 254

"8. Personnel by Discipline"
The rules that go with the questionnaire require that staff be listed only once, thus requiring some hard choices when classifying an "Architect"

who may also be a "Planner: Urban/Regional" or when classifying a "Civil Engineer" who may also be an "Estimator." The best way to answer these choices is to find out—by some strategy research—which disciplines each particular agency considers important to its work. If in doubt, however, the general rule is to scatter your people over as many disciplines as possible, to illustrate organization depth, rather than concentrating on large numbers in a few specialties.

"9. Summary of Professional Services Fees Received"
Many government agencies have policies to spread their work around, so it is best not to pad the estimates of fees from "Direct Federal Contract Work."

"10. Profile of Firm's Project Experience, Last 5 Years"
Many firms incorrectly interpret this as requiring the whole history of the firm. This is inadvisable. The instructions clearly state: "Select and enter . . . *not more than* 30" experience codes. In practice, it is wise to list only those codes that apply to the particular work of this agency. This makes the firm look more specialized, and will invariably make the agency more curious about you. If the code the agency is concerned about is only one among 30 you list, you can be quickly passed over in favor of a more targeted firm.

"11. Project Examples, Last 5 Years"
Here, the same caveat applies. List only experience that is relevant, and don't list it in chronological or size order: Put first in order the projects that are most similar to the work in which the agency is involved. (It is reported that at one Corps of Engineers office, the contracting officer claims he begins all short lists by looking at the first project listed on question 11 of the Form 254. If that one project is not directly relevant to the assignment under consideration, the firm is passed over.)

Form 255

"4. Personnel By Discipline"
Treat this the same as question 8, Form 254 (above).

"5a. Has this Joint-Venture Previously Worked Together?"
An entire marketing book could be written on the subject of joint ventures*, and this is not the place to cover the entire subject. This question on Form 255 is significant in the context of this chapter, however, because it highlights an area of genuine concern among government

---

*A good starting source on the subject is: *Joint Ventures for Architects and Engineers,* by David Dibner (McGraw-Hill).

agencies. The credibility of joint ventures thrown together as marriages of convenience to win a job is poor. Thus, this is a red flag question. The right answer is "Yes." If the answer is "No," the red flag goes up. Whenever a red flag question appears on a questionnaire, the first step is to contact the agency and try to learn how it will react to a "wrong" answer. Such information can often lead to a change in strategy, or a decision not to pursue the project. In this case, if an inexperienced joint-venture still seems the best approach, one way to answer question 5a is with an (*) asterisk and the words: "See question 10." Then, in that space you can explain in detail the rationale behind the proposed team, and thereby—hopefully—survive the automatic rejection of a reviewing officer who uses this question to screen applicants.

The rule about red flag questions is to never knowingly give a "wrong" answer without explaining in detail why the exception makes sense. If you give a reviewer one good reason to hesitate in rejecting you, chances are you will survive the first cut on that question and stay in the running at least until the next round.

"6. Outside Key Consultants/Associates Anticipated for This Project"

As a general rule, prior experience is considered important here. If you have in mind some consultants with whom you have not worked before, consider omitting them here and introducing the subject later, at the interview. As in the discussion on the previous question, a "No" answer here may be a red flag. At the very least, check with the agency before inserting any "no" answers.

"7. Brief Resume of Key Persons, Specialists and Individual Consultants Anticipated for This Project"

By general agreement across many different agencies, this is considered to be the single most important question on Form 255. If you propose the right people here in the right way, you are almost certain of an interview because the agency will look forward to the chance to meet them. If you propose the right people in the wrong way, you may be passed over for the wrong reasons. And if you don't have the right team for the project, you should question whether it is worth submitting at all.

The basic principles for filling out these resumes are the same as are discussed elsewhere in this chapter under the heading "Personnel Biographies." In a word: Customize. The resume for each person should be customized for each Form 255 submitted. Some specific rules follow:

"a. Name and Title"

In general, organization titles are meaningless. Thus, don't say: "John Doe, Principal (or Vice President)." Say instead: "John Doe, Civil Engineer (or Principal Engineer)." Aim the title at the project.

"b. Project Assignment"

The difference between this answer and the title in a. above is often confused. Don't hesitate to be redundant, e. g., "Will be Project Civil Engineer in Charge," or "Will act as Principal Engineer and Client Liaison throughout the project." Make the function make sense in the client's eyes.

"g. Other Experience and Qualifications Relevant to the Proposed Project"

Here, the most effective approach is to give specific examples of the person's involvement in prior projects similar to the assignment under consideration. The very best answers read like this:

"Served as Project Manager for the $10 million XYZ project completed in 14 months on a fast track approach."

"Designed the long-span structures for ABC Project which won the AISC award for innovative economies in steel."

"Job Captain in charge of production for the DEF Dam, involving management of a 100-person project team for 24 months."

etc.

Above all, what the government reviewing officers want to see in this question is that the people proposed have done before what is required to do the assignment under consideration. It is surprisingly easy to say this, if you know the experience of the person being proposed, and you take the time to customize each submittal.

One more important rule about this question. Don't list as "Key Persons" officers of the firm who may be very important to the firm but will have no real role in the project. This is not the place to satisfy anyone's ego. If the result of loading on a lot of peripheral resumes makes the agency doubtful about who will really be running the project, you will lose points.

"8. Work Which Best Illustrates Current Qualifications Relevant to This Project"

Like it or not, there is no better qualification in a bureaucrat's eyes than evidence you have done it all before. Thus, the more specific the answers here the better, just as is discussed under question 11, Form 254, above. When the examples are less than obviously specific, however, it is important to include a phrase or paragraph explaining why the cited project is relevant. Don't be shy about exceeding the blocks of space allotted on the form. State the case fully, so the importance of the project works for you.

"9. All work . . . Currently Being Performed Directly For Federal Agencies"
This question is particularly aimed at preventing firms from getting more than "their share" of federal work, whatever that means. The point here is similar to question 9 on Form 254. It is advisable to keep this list as short as possible. If a job hasn't started yet or is almost complete, don't list it.

"10. Additional Information . . . Supporting Your Firm's Qualifications"
This is the other vital question, together with resumes of key personnel. Treat it exactly like a cover letter, as discussed elsewhere in this chapter. It is a good idea to transmit every Form 255 with its own cover letter, and if that says it all, simply repeat it here. Never by shy about being redundant or repetitive. Many reviewers only scan the sections of the form they consider important. This is the space to say you want the job and why you feel particularly qualified for the assignment. Say it.

In addition to the standard questions of the government forms, there are a number of other questions which typically show up on questionnaires from other private and public clients. Several of these deserve discussion.

Scope of Services.

Many times firms are asked to describe their organizations in questions like this:

Indicate the scope of services provided by your firm.

The author has seen more than one firm answer this question with:

Normal Architectural Services.

It should be obvious why this answer fails to sell. How much better an answer such as the following:

Comprehensive architectural services including Programming; Feasibility Analysis; Site Selection; Master Planning; Building Design; Construction Documentation and Specification; Cost Estimating; Interior Space Planning and Design; Landscape Design; Construction Supervision.

In point of fact there is no longer a "normal" professional service, and even if there were, how can one be sure that the client's definition of "normal" is the same as yours. All professionals must learn today to define their services whenever they are selling.

Consultants

Many questionnaires ask firms—especially architectural firms—to identify outside consultants they regularly engage or would use on the project at hand. It is wise not to be hasty answering this question.

Quite often, in both public and private work the client may have done some research which led it to blacklist certain consultants whose performance it considers unsatisfactory. At other times, particularly in public work, some consultants may be working just as hard as are the prime architects / engineers to get a piece of the job for themselves.

Therefore, the potential client may be moved to exclude from further consideration a firm which, by its answer to this question, ties itself too firmly to certain consultants.

Where possible, a good answer is to stay loose, possibly with a general statement such as:

> We employ outside consultants after detailed evaluation of their suitability to the particular project, including such factors as prior experience in the building type, geographic location, current availability of personnel, etc. Among the consultants who would be considered for this project are:
>
> (List several qualified firms in each consulting area.)
>
> Final selection of consultants will be made only after review with the client.

Assignment of Personnel

One question very often asked goes:

> "Name the key personnel who would serve as project manager, project designer, and job captain if your firm is selected, and give a biographical sketch of each."

This is a favorite question with government agencies and is often also asked by major corporations and other organizations who have had considerable experience dealing with larger professional firms. They know from both happy and unhappy past experience how much the smooth progress of a job will depend on those assigned to its day-to-day development, and they want to be assured in advance that the job will be managed by qualified personnel with whom they can work harmoniously.

Most professionals would agree in principle with all of this, but they have learned from their experience that it is almost impossible to promise three key people to a prospective client at the paperwork stage, since it may be months before a selection is made.

A major quandary occurs when a firm has what are obviously the "right" people on its staff, but they are not immediately available for reassignment.

In such cases, it is entirely proper to answer the question along these lines:

> Key personnel are assigned on the basis of their experience in the project type and their availability at the time the commission is awarded to us. Because we cannot be certain when this project might be awarded to us, if at all, we cannot guarantee indefinitely that specific personnel will remain available for this assignment. Attached are biographies of two project managers, two designers, and two job captains from whom we would *hope* assignments for this project could be made. If we are selected for final consideration for this project we will then, of course, stipulate personnel to be assigned, and if they are unsatisfactory to the client we will withdraw.

This answer, obviously, has two advantages: It does not commit the firm to tie up key people waiting for an indefinite assignment, but it does permit the firm to submit biographies of its strongest personnel, even though it cannot be assured all of them will be available at the time the job is finally awarded.

Personnel Biographies

Clients are focusing more and more during the paperwork stage on the qualifications of the team of people proposed for the project under consideration. In this climate, the firm that knows best how to present the credentials of its people usually makes many more short lists than firms that propose their project organization in a canned or perfunctory way.

In preparing biographies of key personnel for presentations it is essential to tailor them to the assignment being sought. To do this, resumes must be custom rewritten for every project submitted. This can be an enormous amount of work—it is becoming a full-time job in large firms that submit several paperwork presentations a week—but the payoff is worth the effort in every size firm, and the task can become much easier if the same format is used for presenting all biographies.

The most effective formats seem to be those that present the information in outline format, both because it is easy for the reader to focus on desired information, and because it is easier to insert and edit custom sections. A composite of the outline format for biographies used by firms with a lot of experience in marketing their people would include these elements:

Name

Functional title on the proposed project team—*not* the organization title (e.g., "Site Engineer in Charge" or "Project Manager," not "Vice President").

Role in proposed project. This is an optional touch, but where many resumes are being submitted it can greatly reinforce the validity of the proposed team to say for each person something like: "M. _____ will be specially concerned with site and regulatory constraints and will be based in (city) during the entire on-site investigation and programming phases of the project."

Education. Degrees and colleges. (If something specific can add strength, use it: e.g. "Wrote Master's thesis on design of suburban office parks.")

Registration. E.g., "Registered in (location of project) and 22 other states." (If not registered, just don't include this heading.)

Professional Memberships. Ditto.

Related Experience. Here, include one, two, or at most three one-sentence paragraphs describing relevant experience on other projects. These statements must be custom-written for each proposal because there are endless variations in emphasis that can be given to the same experience. For example, one person's project experience could be used in three different biographies as follows:

A) Project captain for feasibility study of the 120-acre XYZ Corporation office campus, which involved site and land planning, land use analysis, and appearance at public hearings to obtain zoning permits.

B) Project designer in development of Office Campus Plan for XYZ Corporation encompassing seven buildings each tailored to a different department, all on a 120-acre site where environmental preservation was a major constraint.

C) Facilities programmer for 1.5 million sq. ft. XYZ Corporation World Headquarters, involving interviews with 100 corporate managers and computer processing of personnel and equipment inventories, all leading to a master space plan for seven different buildings each tailored to the requirements of the occupying division.

Other Activities. Here, mention can be made of civic or professional activities that bring credibility to the assigned role (e.g., "Keynote speaker at Conference on Land Use Planning").

Biographies in this format should never be longer than one typewritten page. Don't oversell, even if you have the ammunition. If one person has dozens of relevant projects, choose only the best two or three so you don't make one person look far superior to others on the team.

In terms of graphics, photographs with the biography can be a nice touch in many presentations. If a photo format is adopted, print the pictures on blank pages and then add the custom text by work processor, photocopy, or offset.

The purpose of custom resume preparation is to make the prospect aware that there has been special effort to put this team together for this project. One key to getting this done efficiently is to orient everyone who may be proposed on a project team so they understand the firm's biography format and can help provide relevant background, and have one writer draft all the biographies for the project so they flow into a persuasive team package.

If, after adopting a format for personnel biographies, you have any questions about its effectiveness, go back to a prospective client (whether you won or lost) and ask how your resumes compared to other proposals the client received. The feedback should help fine tune the resume system and may help you learn how to put together better teams in the first place. Professional services is a people business, and no marketing tool is more valuable than the ability to sell your people effectively.

References

Some prospective clients ask for references. Whether they are asked for or not, however, this matter opens a whole new phase of the business development process.

There are two kinds of personal influence which professionals can use to help them get work.

The most frequently used influence system involves friends or contacts who intercede on one's behalf, more or less for personal reasons. It does not take a book to tell those who have such contacts how to play the game.

For those who cannot or do not choose to go that route, however, there is still a very important place in the business development process for use of references on fully professional terms. The formula is:

1. Use references whose judgment will be particularly respected by the prospective client; and

2. Be sure they commit themselves to the prospect.

Often, references are asked for in the questionnaire; but it is a good idea to offer references whether asked for or not. In doing so, remember it is the quality not the quantity of the references which is most important.

Two or three names are usually sufficient. If there is a choice, preference should go in approximately the following order of priorities:

Satisfied former clients;

Persons in the same field as the prospective client;

Persons of stature in the community.

Whenever references are given to a third party, it is a good idea to forewarn the persons cited so they will not be caught off guard if contacted. Sometimes a phone call is appropriate, but often a simple letter like this will do:

Dr. A. B. Jones
Director
South County Hospital

Dear Dr. Jones:

I have taken the liberty of giving your name as a reference to the North County Hospital Board, which is considering our services.

If you should be contacted we will be most grateful for any good words you can say on our behalf.

Very sincerely,

As a practical matter, references are seldom contacted, even when requested. Therefore, if the references are worth listing it may be worth trying to have the reference go on record before being asked.

The way to do this is to write a letter to each reference along these lines:

Dr. A. B. Jones
Director
South County Hospital

Dear Dr. Jones:

I have taken the liberty of giving your name as a reference to the North County Hospital Board, which is currently considering our services for a new out-patient facility.

We feel particularly qualified to seek the North County commission because it will involve team care concepts such as we pioneered with your help at South County Hospital.

The North County building program is being directed by Mr. C. D. Smith, President, North County Hospital Board, 200 Main Street. Should you care to write him directly, I know he will value your comments about our organization.

Cordially,

In this case, there is a 50–50 chance Dr. Jones will voluntarily write a persuasive letter on the firm's behalf. If he does, chances are equally good that Mr. Smith will be impressed. However, if Dr. Jones chooses not to write and is not asked, nothing is lost.

The average professional firm has several dozen references of this caliber available to it, and in practice they are seldom used. It is good business to ask several key references if their names may be used from time to time. Then don't be shy about asking them to put in a helpful word. They are usually flattered to be asked about a specific case, even once or twice a year.

The Cover Letter

The best executed questionnaire and the best prepared brochures may receive scant attention if they are not accompanied by a strong and effective covering letter.

A simple transmittal referring to the attachments is worse than no transmittal at all. The letter of transmittal must emphasize the firm's specific qualifications for the job, and it must stress the firm's real interest in the assignment.

The following is an example of an ideal transmittal for a custom-prepared questionnaire and two inches of brochures:

Chairman of the Board
XYZ Hospital

Dear _____:

We are honored to have been invited to be considered as architect/engineers for the new XYZ Hospital.

We have learned with particular interest the problems of budget and timetable which your Board faces. We believe we may be especially qualified to help because of our experience with two very similar projects we have completed in the past three years. These are:

ABC Hospital, where we programmed, designed, and supervised construction of a completely new 200-bed institution within a total timetable of 30 months in order to release the original building for urban renewal condemnation.

DEF Hospital, where we designed a 100-bed community hospital at a cost per bed 20% under the state average and assisted the institution in obtaining federal financial assistance for 85% of the cost.

In addition to our direct experience with projects similar to yours, our offices are within walking distance of your new site, and two of our partners have their homes adjoining your property. We can, therefore, assure you the closest personal attention to achieve standards of design that will make the new XYZ Hospital a physical as well as functional asset to the community.

Enclosed are the completed questionnaire and brochures requested in your letter of inquiry. We would very much like the opportunity to meet with your Board so that we may present our experience and interest in person.

Very sincerely yours,

Obviously, not all transmittals can be quite as pointed as this. But it should be clear that a well written transmittal can say in 200 words all that is important in 20,000 words of questionnaires and brochures. Never rely on the client to dig this out of the form work and brochures. Summarizing it succinctly in the transmittal letter is the only way to assure that the message will be duly noted.

Smaller firms, without previous experience in the particular building type, must strive to appear qualified on other grounds, such as educational background, professional committee work, civic activities, etc.

On the other hand, very large, experienced firms, must strive to appear sufficiently interested to reassure the client that his project will receive proper attention.

Several ground rules about transmittal letters are worth noting.

1. The primary purpose of the transmittal letter is to get on the short list. The job getting comes later.

2. The letter is a selling document, not a formal proposal.

3. Brevity is essential. If questionnaires are attached, a letter of one to 1½ pages is long enough. If no questionnaires are involved, and the letter must be the complete statement, 3 or 4 pages should be the outside maximum. (Even in such cases, a one-page transmittal and a three-page attachment may be more effective.)

4. Always conclude the letter with a request for an interview.

Exhibits of Experience

Quite often, the requirements for paperwork submittals include requests for illustrations of related or similar projects executed by the firm. The first question that arises in fulfilling this requirement is whether the firm's standard brochure will suffice.

Sometimes the brochure will do, but there is always the question of whether it is specific enough. If, for example, the application is being made for bridges, the first illustrations which the reader sees should be of bridges. And if the firm's standard brochure illustrates 10 pages of power plants, transit systems, and industrial facilities before the first bridge is shown, the application will suffer. Some brochure systems are flexible enough to permit bridges to be illustrated first whenever desired. (This is discussed more fully in Chapter 17.) But even these are frequently out of date, and do not include the latest work of the firm.

In general, regardless of what is included in the standard brochure, it is helpful to include in the custom paperwork presentation one or several project examples that directly address the interest of the prospective client. Often, they can simply be attached as loose photographs with captions. In more elaborate presentations they may be bound into a loose leaf binder containing the questionnaire and transmittal.

The point here is that customizing a few illustrations to include in the paperwork package is generally a plus, regardless of how excellent is the firm's standard brochure.

Whenever paperwork can be aimed directly at the assignment being sought, and appears to the recipient as something assembled particularly for that job, the message conveyed is stronger than when it looks like boilerplate. This rule should be remembered in selecting the whole graphic format used to present paperwork submittals. Make the basic contents look like you cared enough to prepare it all individually for this one submittal. Then attach the stock brochures to that.

PROPOSALS

The distinction between a qualifications presentation and a proposal is often clouded because they are so frequently combined, but in practice the elements that distinguish a proposal are:

1. A clearly defined approach to the work or scope of services;

2. A price;

3. (Sometimes) A proposed design solution.

The proposal is becoming an increasingly common part of the development process because of the rapid expansion in the nature and scope of professional design services, and the desire of clients to be clearer about what services they will receive before they make a selection.

Many sophisticated marketers of professional services will volunteer a proposal even when it is not requested because they feel confident that what they put on paper will help outclass their competition. In other cases, the professional will conceive a service the client has never thought of and by writing the proposal will actually be designing the job. This sort of thing is quite common in search of federal or foundation grants for research or prototype studies.

The government has used the proposal method of screening competitors in other fields for years, and is extending the practice more and more into architecture and engineering, particularly on unique projects. Often government agencies going this route will first narrow down the list of competitors through interviews, until three or four firms are selected for a final round. These firms are then issued a formal "Request for Proposal" (RFP) which calls for a detailed work statement in reply. RFP's can be several hundred pages in length for a project such as a mass transit system, and by rule of thumb the responding proposal should be about the same length. In such cases, responding firms may be required to assemble a work team to think through the project, prepare critical path schedules, organization charts, and the like, and submit the resulting proposal in multiple printed copies. Such an effort can run into thousands of dollars, and is obviously not entered into lightly. This sort of proposal, of course, is usually called for only on larger or more complex projects, but in principle the ingredients are no different from those which should make up an effective two-page letter proposal.

Defining Services

In essence, a proposal outlines a scope of service and sets a price. When proposals are drafted solely to record a legal agreement, they can be solely technical in nature. However, when a proposal is drafted before a commission has been awarded, it is a selling document and must be written to sell.

One principal difference will be in how the services are defined. Technical jargon, such as "programming," "schematics," "working drawings" can be totally incomprehensible to the layperson. Such prospects may get the drift of what happens in these phases, but quite likely will not understand what it really means to the project, why the service is needed, or why it will cost so much.

For example, this is a perfectly proper way to propose a service:

II. Schematic design of the site plan, roads, utilities, and building, to establish general locations, and massing. Fee due at completion: 15%.

On the other hand, the same services can also be proposed in the following manner:

PHASE II—Schematic Design—In this phase we will translate the program requirements into three-dimensional solutions, covering particularly:

Site access and egress. Determination of best location for vehicle and pedestrian entrances and exits, including particularly an analysis of the traffic light situation on Main Street, and the feasibility of sidewalks.

Roads and walks. Layout of streets and paths to preserve a maximum of the site amenities while keeping costs realistic.

Utility grid. Establishment of a master utility distribution plan for water, sewer, storm drain, electricity, and gas, planned to accommodate not only the initial program but also to allow most economical expansion in the future.

Building placement and form. Development of optimum building area and height in relation to the functional requirements, topography of the site, zoning, aesthetics, and cost, with special emphasis on the relative merits of a single structure versus a multiple-building complex. Alternate possibilities will be explored throughout the Schematic Design Phase and all feasible schemes including relative cost estimates, will be reviewed with the client. The recommended scheme will be developed both in drawings and a three-dimensional block model for presentation to the Board of Directors, together with a Schematic Cost Estimate which we guarantee to be reliable within 10%.

Upon approval of Schematic Design, 15% of the total fee will be payable.

A good proposal makes specific reference throughout to the individual factors of the project involved. And as it describes each service it will define the involvement of the client and will clearly specify the output of each phase.

Proposals are also opportunities to define for yourself how much work is involved in a given project, and thereby more accurately estimate the fee that should be charged.

This principle can even work two ways. A landscape architectural firm faced a difficult pricing problem as it expanded into planning what have come to be called planned unit developments. Its clients, mostly commercial developers, were accustomed to paying fixed per-unit prices for a) subdivision site engineering (street and lot layouts) and b) for landscape plans. The aggregate of these two services would cost about $10,000 for a 100-acre development.

The land planners, in order to design a full-fledged planned unit development, were providing a total of 16 services which the firm valued at $100,000 for a 100-acre development. Its clients, however, were refusing to pay much more than the old $10,000 fee because they did not understand the additional scope of services in quantitative terms.

As an experiment, the land planners rewrote their proposal into an itemized list of 16 "Standard Services:"

Group I   Services—Planning Phases
    1. Regional Land Use and Site Selection Studies
    2. Site Evaluation and Feasibility Study
    3. The Master Plan

    4. Public Approvals
    5. Financing Submittal

Group II Services—Design and Construction Phases
    1. Preliminary Design
    2. Architectural Coordination
    3. Construction Drawings
    4. Inspection

Group III Services—Supplementary
    1. Architectural Coordination
    2. Graphic Design
    3. Presentation Materials
    4. Interior Design
    5. Promotion
    6. Market Surveys

The proposal document describes each service in a complete paragraph and quotes a separate unit fee for each of them. By reading this list of "standard services" the developer client can see everything being offered and how much it will cost. If the client wants to reduce the fee, it becomes obvious that some services have to be deleted.

When this proposal system was introduced by the land planners the results were an average five-fold increase in the fees the firm was able to put under contract for typical projects. More important, the firm found it had a much easier time selling all the services of which it is capable. (The complete document referred to in this illustration is reproduced in the Appendix, not only for the reasons cited, but also as an example of the effectiveness of good clear writing in drafting a proposal.)

Design Solutions

The tendency to require at the proposal stage some sort of design solution is again on the increase at this writing. How long this will remain a factor in marketing design services remains to be seen, but historically the practice has run in cycles of a generation or more in length. In the nineteenth century the primary marketing an architect did was to submit competitive sketches until one caught the eye of an interested client. In engineering in the same century many of the major bridge commissions were awarded to the engineer who submitted a design and said it could be done in the most persuasive manner. The practice of clients requesting free sketches became so onerous by the early twentieth century that one of the major reasons for founding the American Insitute of Architects was its efforts to discourage and limit unrestrained design competitions.

For the first three quarters of the twentieth century the practice of submitting—or being requested to submit—design solutions prior to selection was on a steady decline. The AIA had suggested rules for formal competitions but they were so complicated and cumbersome neither clients nor practitioners had much stomach for them. Then in the mid-1970s, the federal General Services Administration experimented with a selection process in which a short list of four to six firms were paid a nominal stipend to submit design solutions prior to selection. Much to everyone's surprise, architects and their consulting engineers jumped at the chance to enter these selection competitions, and freely spent many times the stipend on their submittals. The client, faced with the chance to see several alternative schemes before being committed, was delighted. Once word got out that architects would play the game—and the separate government antitrust actions eliminated the ability of the professional societies to declare free sketches unethical—the lid came off.

Engineers have been subject to less of this to date, but the trend is evident in the number of major projects where the client will hire one engineer to make preliminary "studies" with the stipulation that the original engineer is excluded from the implementation. Then, when the client is satisfied it has chosen a scheme it wants, it will take competitive proposals from other engineers to carry out that design.

What all this means in fact is that clients are increasingly wary of putting their needs in the hands of a design professional without knowing in advance what they are going to get. When seen in this light, there is an obvious first marketing opportunity in trying to convince the client that the normal services of the firm include exploration of a variety of alternatives and there is no need for a competition just to see options. When this does not work, and the client is committed to seeking competitive design proposals, the marketing approach used to present design solutions can become as important as the solution itself.

Some firms look on a design competition as an opportunity to demonstrate how creative and detailed a scheme they can propose for the assignment. In many competitions this is what is wanted and the only marketing choice is whether or not to play the game. In other cases, however, the presentation of a proposed solution is really an opportunity to demonstrate to the client the design process in which the client would participate if the commission were to go to the proposing firm. In such cases the design solution can embody elements such as processes for evaluating alternatives, employee and or/public involvement, financial options, etc.

Developing proposals that respond to RFP's calling for design solutions is the most costly and highest risk part of the marketing paperwork game. At the very least, design firms must be very cautious before investing much of their marketing budget in this type of proposal. Only when the list of competitors is reasonably short, or the firm's confidence in its potential to match the client's needs and taste is very high, does submitting this kind of proposal make a lot of sense. Meanwhile, keep a wary eye on the marketing cycle: The practice of being asked for design solutions prior to selection may get worse before it gets better.

MANAGING THE PAPERWORK PROCESS

A very important component of the marketing paperwork process is how the material is assembled and the package put together for presentation. A number of different skills are involved, including:

Marketing input about the client and the strategy to be taken;

Technical input where services and approach must be defined;

Management input in selecting the team to make the presentation;

Writing, to say it clearly;

Graphics, to illustrate charts and graphs, and to present it attractively;

Typing, to produce it.

It is rare when all these talents can be found in one person. In a small practice where the principal does almost everything, the paperwork job is often done with canned brochures and boilerplate simply because there is not the time to do each proposal in a custom manner. Such firms need to ask themselves whether they are playing in the wrong leagues if their paperwork batting average is low.

The firms that do well find that the best paperwork is produced when a team is assigned to work on the presentation for each project. Typically, the team will include the person marketing the project, any technical specialists that will be involved in the project, plus a marketing writer, graphic designer or draftsperson who will do the illustrations, and the secretarial staff to produce it.

If this team gets together as soon as the need for the paperwork presentation becomes apparent, and spends an hour or two developing a detailed outline of what will be included and who will assemble each element, the process becomes both manageable and more efficient. *

---

* An excellent monograph on the operation of proposal teams in larger firms is titled *Proposal Preparation* by Diane C. Creel, available from Society for Marketing Professional Services, 1437 Powhatan Street, Alexandria, VA 22314.

Whenever such a team is assembled and finds it difficult to get excited about the effort to prepare the submittal, it is time to take a hard look at whether to pursue the job. Many firms make 50% or more of the short lists for which they submit paperwork. Others, make only one in five or one in ten. The goal should be to keep the ratio as high as possible. Whenever the average falls below 50%, and one is confident the firm is going after the right projects, it is worth looking at the content and quality of the paperwork being submitted.

In conclusion, one final word must be stressed about all the free advice in this chapter. As noted earlier, once every competitor gives the same answer to a question, that answer—no matter how "right"—becomes meaningless. The examples cited in the foregoing material will be useful only if they stimulate readers to seek their own creative way to communicate on paper.

If this chapter has only one big message, it is this: Don't rely on canned material. Those who put time and sincere effort into written submittals, tailored to the problems of the client being sought, will be amply rewarded.

This phase of the business development process concludes at the post office, when the paperwork is put in the mail. (Hand delivery is seldom necessary, except to beat a deadline.)

Now one must wait for a response. During the waiting period, it is perfectly proper to keep in touch with the client if the response is not received on the timetable originally outlined. But don't push or ask for more than the latest timetable. Clients invariably overestimate the speed with which they can conclude this screening process. Delays are the rule, rather than the exception.

Business development is a game in which the postman (or telephone) rings twice—first with the request for submittals; second with either a rejection slip or an appointment for an interview.

The day will eventually come when the postman will bring a return letter from the prospect. It is a rare professional who can open this letter with a wholly steady hand.

from: **the Coxe Letter**

CASE STUDY #94: A COVER LETTER THAT GETS THE CLIENT'S ATTENTION

Last January the Corps of Engineers advertised in CBD for the design of an access road into a national recreation area in Tennessee, along with ancillary parking, visitor facilities and landscaping.

Before submitting a proposal, Gresham & Smith, a 115-person Nashville A/E firm, set out to get more information about the project. Bob Sylar, Director of Civil Engineering, telephoned the Corps' project administrator. He said, "We'd like to know as much as possible about your project so we can decide how to put together our proposal."

His research produced some detailed information on the scope of work and expertise needed. When Sylar shared it with partner-in-charge Fleming Smith and marketing staff Jerry Guy and Orlean Gibson, a proposal strategy emerged. They concluded that structuring a team of experts was vital, and that the team needed to include a landscape architect, structural engineer, geotechnical engineer, planner, and—deciding that the project's bridge (not played up in the CBD ad) was key—a bridge expert. The team they put together included outside consultants as well as appropriate staff members.

Once the team and strategy were identified, Marketing Coordinator Orlean Gibson wrote the proposal.

The cover letter of the firm's response is reproduced below with only the names of consultants and staff deleted. Marketing Director Jerry Guy comments, "This is one of several styles we use for cover letters. In this case, we emphasized the team." Regarding the dual-signatures of partner-in-charge and project coordinator, Guy notes: "We think it indicates a special commitment on the part of top-level management."

The Corps received 75 responses to its CBD ad. After review, the Corps said Gresham & Smith's proposal was "by far the best", and the firm was ultimately awarded the job at roughly a $½ million fee.

Nashville District Corps of Engineers, Nashville, TN 37202
ATTN: Engineering Division, Relocations Branch

L _____ ROAD AND BRIDGES
B _____ NATIONAL RIVER AND RECREATION AREA

We are pleased to respond to the COMMERCE BUSINESS DAILY announcement of the referenced project. We enclose our SF 254, SF 255, and SF 254's for each of our predetermined associates. As you review these forms, we would like you to consider the following points which make us believe the team we've assembled is the most qualiffied to do the job:

1. Each of our predetermined associates has been chosen for a specific area of contribution to the total team effort:
   A. XXX ENGINEERING CO., INC. was chosen for their specific expertise in the area of field surveys. Team members from XXX will work closely with Gresham and Smith surveyor, XXX, to insure project coordination among all team participants.
   B. XXX ENGINEERS, INC. will have as their primary team responsibility providing structural expertise particularly in the area of bridge design. As you will note from their SF 254 and from the representative projects listed under question 10 of

our SF 255, XXX has had extensive experience in a variety of bridges and has been recognized by several groups for their innovative designs. Another consideration in our selection of XXX as an associate is the fact that their firm is located in nearby XXX, Kentucky. We believe that concentrating our team members in locations geographically accessible to each other and to the referenced project will enhance project supervision and coordination.

C. XXX ENGINEERING CO., INC., a Nashville-based firm, specializes in the collection and analysis of subsurface data and in design input concerning choice of paving systems and materials. XXX has done extensive work as a consultant to Gresham and Smith in soil testing and related subsurface investigations so they have the added advantage of being thoroughly acquainted with our design team approach.

2. As the primary firm, GRESHAM AND SMITH bring to the project a multidiscipline team with a decade of experience in a wide variety of projects. Our projects have spanned the United States and several foreign countries and, as a consequence, our design team approach has become highly refined. We are accustomed to coordinating input from multiple team members and have developed a number of techniques to insure that all projects meet time and budgetary constraints:

—As partner-in-charge, I will monitor the quality and efficiency of our service to the Corps of Engineers. In particular, I will be involved in the initial phases of project planning to insure that instruction received from the Corps is interpreted properly.

—XXX will serve as total project coordinator and will hold regularly scheduled team meetings to discuss all aspects of project delivery. Of XXX's 20 years experience in civil engineering, some 14 years have been spend in the design and construction of roads from rural farm to market to sophisticated interstate systems. In addition, XXX will direct the efforts of our other four in-house civil engineers all of whom have had personal experience in road design.

—Gresham and Smith architectural team members, XXX, XXX, and XXX, will combine their architectural expertise and sense of aesthetics with bridge experience to produce architectural alternatives for the crossing of the Big South Fork.

—XXX, XXX, and XXX will, respectively, coordinate cost control, landscape, and planning aspects of the project for the Gresham and Smith design team.

Finally, we have a genuine desire to serve the Corps of Engineers, on this and future projects, and we look forward to the opportunity to do so.

GRESHAM AND SMITH

Fleming W. Smith, Jr.
Partner

Robert L. Sylar, P.E.
Director of Civil Engineering

CASE STUDY #98: HOW TO WRITE A 255

Some time ago Hathorn Associates of Jackson, MS, submitted a 255 form for a V.A. hospital alternations project in Biloxi. The firm didn't make the shortlist, and partner Gary Hathorn decided to find out why.

"I called the V.A. and they said if I made an appointment, a member of the selection committee would go over the 255 with me. I went to Washington and visited the vice-chairman Roland Vaughn. He gave me half an hour and went through the 255 item by item.

"He said the 255 doesn't give A/E's enough latitude in describing their experience." Vaughn suggested adding a page to supplement the resumes to describe past project experience that is similar to the project at hand. On Question 8 ("work . . . which best illustrates current qualifications relevant to this project") Vaughn said, "We'd like to know more about each one of them."

Two months later Hathorn responded to another CBD announcement, this time for a $7 million-plus surgical facilities addition to the same Biloxi hospital.

"I put together a much different 255, with a great deal more detail. We rewrote everything." (It took about three times as long as the previous 255.) "On the resumes, we named the hospital, its cost, and the person's responsibility on the project. After Question 8, we inserted several pages with a couple of short paragraphs on each of ten projects—describing the size of the hospital, what kinds of facilities it included, the scope of our work, and so on."

Hathorn's 12-person architectural firm was shortlisted and eventually won the job. Hathorn concludes, "We also do a lot of work for the Corps of Engineers and the Navy. In the future, I'll go into more detail with them, too. You just have to tell them more about yourself."

CASE STUDY #100: SMALLER FIRM MAKES 95% OF SHORT LISTS

As a two-man design office founded in 1976, Lyon & Brown, Inc. of Redwood City, CA often found itself in competition against firms of 40–50 employees. One of the resultant problems was a perceived skepticism on the part of prospects concerning the firm's capacity to handle an increased workload. The firm's solution was to include a meticulously detailed project schedule—which includes a graphic representation of all work currently being done in the office—with each proposal. Since instituting the technique last year (as well as becoming more selective in its marketing goals), the firm has been shortlisted for virtually every proposal.

Lyon & Brown also includes a CPM schedule which illustrates how the prospect's project would interface with the office's present job schedule.

"Naturally, we had alluded to schedule in the past," explains Vice President Troy J. Brown, Jr. "But we've always done what most people are doing in their proposals: verbally assuring the client, in effect, that we can make his schedule.

Now, instead of promising him we can do it, we're demonstrating how—graphically.''

Another part of most of the firm's proposal packages is a schedule of costs, broken into about two dozen catagories. The cost information provides the prospect with ''something more than just the feeling that the fee is a little heavy,'' says Brown. ''They have something to work from.''

The firm recently used its proposal package in applying for a commission from the Church of Religious Science—a project type for which Lyon & Brown had no prior experience, though the competition did. Besides cost- and time-consciousness, another point stressed by the firm was that, as designers with no preconceived notions about church design, the client would be assured of getting something more than a recapitulation of old ideas. Lyon & Brown was awarded the commission to design the $650,000 church.

13

COMMUNICATING
IN PERSON
—THE INTERVIEW

"Make simple presentations. Don't talk about what you want to do, but what you can do."
Client Donald P. Simek, IBM

Interviews are exercises in human chemistry.

At this stage in the selling process, facts are much less important than the personal interaction of the client and the professional. Every professional can tell when they meet a client with whom they would really like to work. The client reads the professional in just the same way. This, then, is the purpose of the interview—to convince the client that working with you will be an interesting and rewarding experience.

Interview tactics are quite different from those of the written submittal. The purpose of the written material was to encourage an intellectual reaction—to sell the firm on substance. In the interview, on the other hand, the objective is to encourage a human interaction.

Also the risk/reward ratio changes at the interview. During the long list stage, when the primary object is to make the short list, the odds favor conservative, low risk approaches, which give the client the fewest reasons to reject you. There are a number of places on the short list, and landing any one will do. At the interview stage there is only one winner. This, then, is the time to take higher risks and go for broke because the odds are, if you play it safe, you will lose.

The starting point of the preparation for each interview is to learn as much as possible about the conditions under which it will be held, particularly:

Who will be interviewing—how many people will be present; what are their names?

The physical setting—can slides be shown; will there be a screen or should the firm bring its own; are there easels for charts; etc.

The timetable—is this the first or last interview of the day; how long will it last?

The agenda—are there particular subjects the interviewers would like to see or hear about?

The best source of this information is a phone call to the person scheduling the interview. Normally the client will be glad to be helpful, even in cases where earlier strategy research information has been hard to get. By the time two or three dozen firms have been screened down to a handful for interviews it is likely the client will have become personally interested in each of the remaining candidates and will be anxious to help each firm sell itself.

(There are unpleasant exceptions, of course. Some time ago a Pennsylvania school board invited 22 firms to interviews, without any prior screening whatsoever. Each firm was allotted 15 minutes, and the interviewing committee ran them through like cattle—a timekeeper cut off the proceedings in mid-sentence whenever fourteen minutes and thirty seconds had elapsed. But clients who treat interviews in this manner are saying a good deal about themselves and the kind of projects they want to build, and the professional is free to make a judgment whether the firm wants that kind of work.)

The ground rules for the interview will have a major influence on the approach and strategy to be followed.

In the first place, the ground rules will define whether the interview will be formal or informal.

Sometimes the interviewing client will be a single individual acting like the patron of old with complete authority to define the criteria, establish the terms of the working relationship, and make the selection. Developers, corporate entrepreneurs, and small businesspeople frequently operate in this manner. Their interviews are totally unstructured and generally result from a simple invitation to "come over and talk." Even though the client may be accompanied by an aide or two, they alone will make the decision. This format usually calls for an informal interview aimed at establishing rapport with the single decisionmaker.

More often, the interviewers are a committee of a public agency or private organization to whom has been delegated the responsibility for selection. Frequently such committees will have little or nothing to do with the subsequent execution of the job. Sometimes the interviewing group is merely a screening body with power only to recommend professionals for selection by higher authority. It also makes a very great difference whether or not the interviewing group includes a qualified architect or engineer. Whatever their makeup, such groups tend to establish fixed formats and schedules for their interviews and they often issue invitations that stipulate the topics to be covered. These situations, by definition, are formal interviews.

Each case calls for its own strategy.

INFORMAL INTERVIEWS

The one common denominator of all interviews is the architect or engineer's desire to get the client interested in the professional as an individual. This, of course, is easy to say. But how does one do it?

To start with, a certain amount of self-confidence is very helpful. This can be obtained by practice (e.g., going to a lot of interviews); by taking courses in public speaking and effective selling; or by volunteering to serve on a selection committee and watching how one's competitors do it.

Beyond self-confidence, however, there are some tactics which can do much to make an interview the two-way interaction it should be.

First of all, in informal interviews try to avoid making a direct presentation of your case at the outset. If you are one of several professionals the client is talking to it is quite likely the client has heard much of the same material before. And if you are the only firm being talked to, the client may not yet know why to care about the information being presented. In either case, a direct presentation should be avoided until the client has become interested in you.

To do this, start off the interview by asking questions of the client.

It is perfectly good form to open an interview with a statement like this:

I/we have come prepared to tell you about how our firm operates, show you some of our work, and explain why we feel qualified to serve your needs. Before going into our presentation, however, there are a few questions we would like to ask to help us know a little more about your requirements.

Normally, the interviewer will be delighted. And as the client learns, by the professional's questions, that you care about the problem to be addressed, the client will begin to care about you.

Appropriate questions are related to the same programmatic data discussed under strategy research in Chapter 10. If the professional has done this earlier research successfully, there will be many constructive questions to ask. Examples might be:

Has the program been translated into a detailed space analysis?

How was the size of the site determined?

Have you determined what zoning clearances will be needed?

When will the site be available (cleared) for construction to start?

How do you expect to finance the project?

Is financing available now, or will there be a wait after design while money is raised?

Have you determined a philosophy of how this building should function?

Have you considered how you will pick your builder?

How will you be organized to direct the project?

Who will be in charge of liaison between yourselves and the professionals you select?

Etc.

Often, these questions will open avenues for the professional to mention specific services that can be provided to solve the problems under discussion. More importantly, the questions will frequently encourage the prospect(s) to begin talking. Quite often it will develop that many of these same questions have been on the mind of the client but have not been openly discussed before.

The lucky professional who sets off such a dialogue will be scoring points faster than ever could be registered by a formal presentation.

If these well-chosen questions have done their job, the professional will feel much more comfortable when the time comes to begin the direct presentation. Ideally, a good deal will have been learned about the particular concerns of the client, and the subsequent presentation can be tailored to concentrate on these areas. And the professional will have learned something about the personality of the client, which will help structure the interaction being sought.

Sometimes, the professional's questions will set off a discussion that will use up most of the time allotted for the entire interview. Don't panic. This may be a high compliment, and may well make that interview the most memorable of all to the client.

If time runs short, ask what the prospect would most like to see or hear in the time remaining, and limit the presentation accordingly.

FORMAL INTERVIEWS

The formal interview requires an entirely different organization strategy. In the first place the setting tends to require a direct presentation at the outset leading, hopefully, to some dialogue at the end. Secondly, an interviewing committee involves a number of different personalities, and the human connection is much harder to achieve. (Most committees, in fact, act very impersonally, which doesn't help the professional process at all.) Finally, committees generally make decisions differently from individual patrons, and it is important to recognize this in one's interview strategy.

Committees most often make decisions by consensus, derived from largely objective criteria, rather than by simple confidence or instinct. In this sort of decision process, firms can easily get eliminated for minor reasons simply because the committee starts by looking for reasons to discard a contender. Sometimes if each person on the committee has a different first choice, the selection will go to everyone's second choice by process of elimination. This isn't to say this is the best way to make group decisions. It just happens that way when poorly led committees get involved in selecting professionals and become preoccupied with trying to be "fair" rather than with selecting the best firm.

All this does not contradict the previous statement about interviews being a time for higher risks. Playing interviews so safe that one becomes everyone's second choice may, on occasion, win, but it is a poor strategy over the long term. It is better to go for first place all the time, and take one's chances on the occasions when it doesn't work.

There are five basic questions which should be answered in order to design an effective formal presentation.

What do we have to say to fulfill the formal selection criteria?

Who on our team will best relate to the personalities of the clients?

How should our people project themselves in the presentation?

What media will make it easiest for us to project our style?

What is the one point we want the prospect to remember 10 minutes after we have left the room?

The last question is the most important. If, after making your presentation, there is no one point the committee can remember about you, they will have no point of reference when it comes to comparing your firm to the others in contention. Consider the dynamics of a committee reviewing all it has heard at several days of interviews. Firm names may be confused, and person-alities transposed. What will be used to identify the different contenders will be comments like:

"The firm that did the XYZ job."

"The firm that said they could start tomorrow."

"The one who quoted the lowest price . . ."

"The firm who told us . . ."

Etc.

There is no substitute for coming out of an interview having left behind the best one point. The point may be specific, "They said they could do it quicker than anyone else", or it may be general, "They are clearly more professional." Whatever point you chose on which to stake your case, this is the place to take some risk. For if your one point is the same as everyone else—"We have a terrific record at cost control and meeting budgets"—it will be no point at all. Playing it too safe on the choice of your one point runs the greatest risk of choosing a point that others mention. And when all firms give the same answer to the same issue, no matter how correct, the issue will become meaningless in the final decision.

The route to selecting the one point to make in the formal interview must come from the strategy process cited earlier in Chapter 10.

Strategy Research	=	Strategy Judgments	=	Interview Message
What do we know about the client, program, site, etc.?		What do we decide about what we know? What benefits do we bring to this assignment?		What do we say about each issue and what is the one point we want to be remembered by?

No handbook can tell you what is the "right" interview message. This must come from your confidence in your abilities, your assessment of the needs of the client, and your response to those needs in terms of the approach and direction you would take if commissioned to do the job. In a very real sense this is why the client bothers to interview more than one firm for a job—to find out what each firm might do with the assignment. There was a time years ago when most interviews were recitations of firm experience and qualifications. No more. In the new competitive environment between professionals, the client knows going in to the interviews that most if not all the firms being considered are qualified to do it. The interview selection is therefore decided more on issues related to the project itself than anything else.

A member of an interview committee of lay people, who were considering firms for an institutional project, recently summed it all up succinctly when she said, in discussing two contenders: "I felt (firm X) just wanted the job; I felt (firm Y) was really interested in helping us!"

The First Five Minutes

The process of planning for formal interviews should be as formal as the presentation itself. All professionals experienced at marketing their services become, at times, self-confident about their interview "formula" and are tempted when they are busy to "wing this one." Don't, if you really want the job. Formal interviews should be strategized at a planning session involving everyone on the marketing team; the roles and inputs should be carefully assembled into a written outline; and there should be a rehearsal. In planning the outline, there are a number of tactical points that deserve special consideration.

What happens during the first few minutes in the interview room can be crucial to the success of a presentation. The objective is to get the attention of the members of the committee so they will be alert (and even awake) for the meat of what is to be said. But frequently, the suggested agenda of what the committee wants to hear—e.g., background of the firm, organization for the project, etc.—leads off with the least stimulating information. Therefore, it is worth giving special attention to the first five minutes to be sure to give the committee some reasons to pay attention through the boilerplate that is to follow.

How the team is introduced is the first consideration. If the professional group comes in and just gives everyone's name, there is little reason for committee members to remember who is who. It is better for the moderator (or each person, if introducing themselves) to say something like:

This is _____ who will be the chief structural engineer on this project.

I am _____ , a professional planner, and my role will be to get the zoning approved.

_____ is our chief of specifications and is the one who will write the 300 pages of conditions to see you get exactly what is called for no matter who bids the project.

Etc.

In addition to identifying each person's role in the project, it is sometimes proper for the moderator to say something personal about each member of the team. These comments can range from the very formal to very intimate, depending on the individuals and the strategy, such as:

This is _____ whom we have chosen as project manager because he is currently serving on the long Span Structure Committee of NSPE and is the most knowledgeable person in this state about the type of project you are undertaking.

or

This is _____ who will be project manager. _____ has just finished a two year assignment on the ABC job, which is very similar to your program. I'm not supposed to say it, but this is the first time _____ has been to an interview and I know she is having a real case of stage fright. I hope you will all help her if she stutters a bit in making her part of the presentation.

Often, a helpful device in introducing the presentation team is to type the names and roles of those attending on a copy of the firm's letterhead and hand it to each member of the interviewing committee.

Another device to get the first five minutes off to a good start is to set forth the agenda for the presentation that will follow. The sage advice: "Tell them what you are going to tell them; tell them; tell them what you told them," is especially apt for interviews. A typewritten copy of the agenda you plan to follow may also be handed out, but be very brief. (Never give a committee something to read when you want them to listen to you. All the lengthy data and brochures that you want to leave behind (so they may read it while your competitors are in the room) should be introduced with a comment like: "I have here our (brochure, project report on X, etc.) and I will leave it with you to review later. . . .")

The most important thing to convey in the first five minutes is some reason why the committee should care about the rest of your presentation. If you are a spellbinding orator, you may capture their attention on charisma alone. If you are a more typical, serious professional, you may need to make your claim for attention on some aspect of your preparation, or your one point. Saying something like the following in the first minute or two may be very helpful:

> We've just completed a study of 32 projects similar to yours built across the country, and later in our presentation we'd like to share with you what we think we've learned that will benefit your project.

or

> I spent last Thursday walking around on your property and reviewing the zoning maps in City Hall, and I've prepared some diagrams to show you later that I think will interest you

or even

> We know the hour is late and you have been listening to a lot of other professionals who have covered much of the same material you have asked us to discuss. We thought therefore we would be very brief, and will take only 10 of the 30 minutes for our presentation, and let you ask questions if there is anything we pass over too quickly.

Get the committee's attention in the first five minutes and your interview will come alive with the kind of chemistry good interviews are all about.

Who Should Attend

Clients are asking to meet at the interview the people who will actually work on their job—not just the marketing principals.

The question of bringing a team to the interview raises the whole matter of how many people should attend. The rule of thumb is: the fewer the better. How many that means depends on the circumstances, but should be governed by insistence that all those who do attend have a meaningful contribution to make. Nothing is worse than taking along and introducing a key technician at an interview, and then having the person sit mute until it is time to leave.

If the interviewing committee requests the entire team be present, or if the firm has specialists it believes will strengthen its case, by all means take them along, but be certain in advance that they have a role to play. And that role should be greater than a simple speaking part in the direct presentation. Every member of the firm at the interview should be equally prepared to participate in the questions and answers and in the open discussion.

When a team goes, another question to deal with is who will be in charge. Frequently a name principal or founder wants to go along to show interest, but must be careful not to so dominate the presentation as to undercut the rest of the team. A good policy is to have the interview managed and moderated by the one in the firm who will be in charge of the client during the execution of the project. If this is a project manager, then let the project manager take charge and introduce everyone else. If there is to be a principal in charge, let that person make clear what the role means by taking charge of the interview. This is particularly important in joint ventures, because, if no one takes clear charge at the interview, the client may wonder who will later be in charge of the project.

Presenting Past Work

For many years the most common materials used at professional interviews, especially of architects, were pictures of past projects the firm had done. Then, some clients began suggesting or insisting that firms not bring slides of past work to the presentation. The reason for this shift is not that clients have become less visually oriented. Rather, it has to do with their disaffection for the way too many firms presented their past work.

The prospective client hires an architect or engineer to solve a problem. Above all else, the client wants to know that the professional will understand the problem and will design a structure to solve it.

Consequently, it is very unwise to base a graphic presentation on the "pretty picture" approach. It is far more important to explain to the prospect how a few selected problems were solved than it is to show a great many different facades or interiors, no matter how attractive if architecture or how massive if engineering.

A "problem solving" presentation may begin with an effective illustration of the completed project, or a picture of the original site, accompanied by a sequence of illustrations showing how the solution works.

If the problem was functional, the pertinent plans and sections of the solution may be shown. If the problem was one of site, it may be appropriate to show views of four or five schematic alternatives which were explored before the solution was found. (Never miss an opportunity to show a committee that there may be more than one solution to a problem. This will reassure them, particularly if some members have pet ideas of their own which they want considered when the project is under design.)

If the problem being illustrated was one of cost, or budget, show the elements which made economies possible. If the solution involves unusual structural systems or construction techniques, pictures of the construction in progress may be very interesting to the committee.

The basic rule is not to show anything which cannot be presented in terms of how it helped solve a problem.

In architectural interviews, it is rarely wise to tell an interview audience what a particular building looks like, or to explain the aesthetic reasons why a particular material or detail was used. Prospects will make their own aesthetic judgments from what they see in the pictures of the professional's work. If they don't particularly like it, and the professional is explaining how beautiful it is, nothing is gained. If they do like it, and the professional is explaining how much better it might have been, all could be lost. Clients rarely hire architects just to design beautiful buildings. First and foremost, the client wants a problem solved. The degree to which beauty should be a factor in the solution will become evident in many other ways, and is best dealt with after the professional has been hired. This statement is not made to advocate or oppose the aspirations of "design oriented" professionals. They have every right to be dedicated to making every project a major design achievement. But the time to accomplish this is during the execution of the project, not while trying to land the commission. The author believes that misunderstanding of this basic axiom is the single most important reason why so many excellent designers have such difficulty in getting work.

Other Interview Graphics

In addition to or in place of pictures of previous work, many interview presentations include charts and other graphics to explain how the project will be handled, illustrate the site, show off the firm's offices, etc.

All these may be effective, particularly if the firm has a strong philosophy of how it processes a project, and can express this in graphic form.

At the same time, routine charts of how a project is developed through programming, schematics, etc. may be repetitive of what the committee knows or other firms have already shown. This sort of material should only be used after asking the committee if it wants to review such elementary information.

The best graphics in every presentation are those that are pertinent to the client's particular interest. Slides of the site, or flip chart diagramatics of the zoning constraints will always get attention. Program representations through block diagrams are another device.

The work of design professionals is graphic by nature, and using visual aids in interview presentations is therefore particularly appropriate. What to illustrate and show has to be guided by the issues to be addressed and the graphic talents of the professionals making the presentation.

The choice of visual medium is a separate matter. There is continuing experimentation going on at this writing which indicates that one should always keep alert to new media. The following generalizations combine the experience of most firms:

"Live" Graphics on flip charts, easels, and blackboards, where the presenter writes and draws the points being made, on the spot, have two advantages no other media provide: The interview audience can be made to feel the material being illustrated is explicitly for them, and not canned; and the skills of the presenter at direct communication of concepts and ideas is clearly demonstrated. Live presentations work particularly well before smaller groups in situations that allow visual give and take—as in illustrating answers to questions about different concepts or details.

Videotape and/or Custom Sound Film. The advent of portable videotape outfits and practical 8mm sound film systems allows the preparation of custom audiovisuals that relate to the particular subject of an interview—such as a site survey, clips of statements by team members who cannot (or should not) attend a presentation, etc. These custom media seem to have a good impact on selection committees, but be careful that the medium does not become more important than the message. These media will probably remain useful only in a minority of special-case interviews.

Prepared Films and/or Sound Synchronized Slide Presentations. Design firms have tried for years to find effective ways to use professionally prepared sound/music/narrated films and slide presentations at interviews. To date the great majority of firms have found it doesn't work particularly well. The material has to be canned, and becomes nothing more than an audiovisual brochure. As one client said, the day after an interview, to a professional who had spent $30,000 and 6 months on a 30-minute movie of the firm's capabilities: "We all thought it was very interesting, but the chairman wondered, when you left the room, who would be seeing it tomorrow night!" If canned visuals have any use at all, it may be as orientation programs in your own office, where they can be shown to touring selection committees (and are a particularly effective way to orient new employees). But that is different from a direct interview tool.

Models. These three-dimensional objects can electrify an interview. They generally require the selection committee to leave their chairs and walk to the model to see it. A model of a previous project that explains unusual concepts can get people really involved in understanding a technical point. A model of a proposed site or existing conditions can become the centerpiece for considerable discussion of alternatives. The disadvantages of models is that they are cumbersome to carry to an interview location, and a specially prepared model may come across as overkill. Nevertheless, it can be extremely effective.

Slides and Photos. Loose slides or pictures boards that can be arranged to suit the circumstances and are presented live by someone on the team are still very effective if, as mentioned above, the commentary makes a relevant point. These will probably continue to be the dominant media for interviews for some time.

The Question Period

At some stage in the interview, usually following the professional's direct presentation, the client will have a list of questions to ask. As a rule, if a questionnaire was used in the written submittal stage, there will be fewer questions at the interview. If no advance questionnaire was used, the client may have a long list at the interview, and may expect to devote the majority of the allotted time to them.

Answering these questions is of considerable importance, although many of them may be unintentionally ill-chosen, inappropriate, or meaningless. Some committees establish a rating system in advance and assign points to the applicants on the basis of the answers given. The professional is advised to treat the questions seriously regardless of how unimportant they may sound.

At the same time, it is good form to avoid lengthy answers which the committee may have heard several times already. For example, to the question: How will you organize to do this project?, some firms will give detailed explanations of every technical specialty and department that will be involved, and illustrate the answer with a complex diagram showing how the team will be structured. While this may be the literal answer to the question and it may be necessary to submit an organization chart, the verbal answer to the question may only need to be: "We use the project manager system and _____ here will be your day-to-day contact." When the committee has seen four or five other organization charts explained in exhaustive detail, you may not need to dwell on the question at all to answer it properly.

The most important thing to remember during a question period is that the questions being asked of you will very likely be asked of all the other contenders. This is because, in dynamics of a selection committee, each member becomes concerned with making their own mark. As the interviews proceed each member will tend to find his or her "own" questions which are different from those being asked by others on the committee. The questioner then "owns" those issues, asks the same questions of every contender, and may make their own selection based largely on their reaction to the answers.

Thus, if a question is technical, it is wise to avoid being specific unless you are professionally committed to your ground. For example, to the question: "Should this be a steel or concrete structure?," the best answer is: "That will be one of the first issues we will want to study for you. We would go about those studies by . . . etc."

Also, beware if the question and answer dialogue causes the committee to ask questions of professional judgment. For example, a discussion of the site may lead a committee member to ask whether the professional considers the property adequate. A direct answer could be very dangerous, no matter what the professional may actually believe, because it could lead to controversy. A better reply may be a statement that a site feasibility analysis would be one of the first services rendered by the firm, if chosen, and concerns about adequacy of the site would be brought to the client's attention immediately.

Similarly, questions about design philosophy should be answered in terms of how a design is evolved, not in terms of adherence to one or another style.

Other questions require commitments, on matters like personnel, fee, or schedule. For these, one must be direct, but need not at the same time hang one's self on a "wrong" answer. An example is the oft-asked question: "Can you design the project so it will be ready for use by (date)?"

Every professional knows how impossible it is to guarantee schedules this early in a job. But it is a very weak answer to give the interviewing committee a lecture on how many things can go wrong with the job which are beyond the professional's control.

A far better answer is to cite other specific projects where the firm has met similar or more acute timetables. Such an answer lets the interviewers draw their own conclusions, and all they really want to know is that their timetable is not impossible.

The rule about question periods is to try to encourage the dialogue, and be very professional in your answers. Try not to get lured out on a limb where the only choices are to jump off or fall off.

The Exit

The conclusion of the interview is usually a bit awkward, for time has run out, others are waiting their turn, and the committee will be a little embarrassed to have to acknowledge that such an interesting, professional discussion has also been a competitive enterprise.

It is important to leave quickly and gracefully when time is up. In fact it is good practice for the professional to be the timekeeper, and when the allotted time is over, offer to leave. If the committee wants to run overtime, it should be entirely their doing.

But do not leave without covering these three points:

1. Ask when the committee expects to reach a decision.

2. Ask if the committee will have the opportunity to visit completed projects before reaching a decision, and urge them to see some of the firm's work.

3. Invite the committee to visit the firm's offices.

Of course, if the answer to the first question is that the committee will vote after the last interview that evening, the only useful move is to go home and pray. But this is rare. If the committee indicates that several days or weeks remain before a decision will be reached or if formal negotiations lie ahead, it is quite possible that the most important chapter in the selling process is yet to be written.

from: **the Coxe Letter**

CASE STUDY #13: CLIENT'S-EYE VIEW OF SEVEN INTERVIEWS

The University of Colorado interviewed seven major architectural teams in order to select a designer for a $6.5 million 13,000-seat events/conference center. The interviews were videotaped by students from the architectural school, and present an extraordinary opportunity to review the interview techniques of a group of design-oriented firms. (Names of the firms, although of public record, are being ommitted by the Coxe Letter, because the winning firm was subsequently unable to consummate its proposed joint venture and resigned the commission. This ironic and embarrassing turn of events perhaps deserves a separate case study but in no way negates the lessons that can be learned from watching the original videotapes).

Colorado's selection process was delegated to a nine-member committee of users, faculty, and students, chaired by William E. Taber, an architect and Director of Facilities Planning. After reviewing 29 paperwork submittals, mostly joint ventures of Colorado architects teamed with national design firms, seven organizations were invited for personal interviews. Each firm was allowed an hour (30 minutes for presentation, 15 for discussion, and 15 for setup and exit) and were specifically asked to address their proposed organization of the project, design philosophy, reaction to the functional program, and to show examples of "projects of similar scope and magnitude."

Recently the videotapes were replayed by Bill Taber for the editor of the Coxe Letter. The seven hour experience was an excellent reminder of what a committee goes through to play the selection game by the profession's rules ("At the end we were kind of wrung out," Taber reports). Some object lessons from the client's point of view:

- Most firms took the exercise entirely too formally, giving their presentations "all the sparkle of a legislative hearing," Taber notes.
- Long presentations don't hold attention. "I had feedback beforehand that a half hour wasn't enough time, but to the committee doing this all day, a half hour comes to be a very long time."
- Design philosophy tends to be the most difficult subject to address. One architect noted that it "tends to be a verbose topic," then spent 11 minutes proving it. The committee really listened only when the architects addressed the design relationship of the project to its site and the overall campus.
- Having too many people present (some brought as many as eight) made the committee "fear that the project would get lost."
- Make slide presentations relevant to the problem. Projects shown to the committee ranged from a $300 million office building to a newspaper plant.
- Being too expert in the building type can made the committee "fear that you already have all the answers—and might be a dictator."

The winning interview was presented by a four-man team that included a local architect, local structural engineer, the design partner of an out-of-state firm with prior related experience, and an acoustical consultant. Highlights of what was clearly the outstanding presentation:

- The team made an immediate positive committment to execute the project from a joint venture office in Boulder, within sight of the site.
- The team introduced a West Coast arena manager as a special consultant and, since he was unable to be present, played a 3-minute tape recording "to infect you with his enthusiasm."
- The acoustical consultant gave a spell-binding exposition of technical considerations that held the committee's attention for 11 minutes (four times the length that had been planned).
- The slide presentation by the design architect covered a large number of projects very rapidly. The commentary addressed program considerations ("This is what the building does . . ." "The essence of the design was . . .").

Following the interviews the selection committee met twice before recommending a first, second, and third choice to the University Regents. (The second choice firm is now doing the job). The obvious common denominator of those at the head of the class was enthusiasm for the project and the ability to create rapport with the committee. Bill Taber puts it another way: "Certain firms came across with an obvious willingness to explore with us."

CASE STUDY #17: A "WOW!" INTERVIEW

The ABR Partnership was chosen from among 37 firms to conduct a feasibility study for an interim location for Denver's Museum for Children. By all accounts, the ABR interview was as much fun as the museum is intended to be fun for kids.

A press release issued by the Museum Board to interested Denver area architects started the selection process. Respondents were given a background questionnaire and asked for a written submission of their interest in the specific problem. A screening committee then narrowed the list to 10 firms for personal presentations and interviews were scheduled in the architects' offices.

At ABR, organization of the presentation was assigned to a 12-member team that would work on the project if they were selected. This team spent one afternoon putting together two different presentation elements:

- A handout booklet for each member of the selection committee describing how the firm would approach the problem. Each page showed (with "fun" graphics) one of the steps they proposed (traveling exhibits, volunteer shows, etc.) to involve the community in determining what and where the museum should be.

- A "learning tree" constructed in the office meeting room of sticks and papers hung from the ceiling contained key words and graphic ideas addressed to what a children's museum might be.

The interview itself was conducted on a Saturday. When the committee arrived almost the entire 30-man staff was on hand (voluntarily). After being shown around the offices the entire group—staff and museum people—gathered informally on cushions under the "learning tree." Wine and cheese were served as each team member described the role he would fill on the project, and the group discussed what they wanted the museum to achieve.

The Museum Committee decision selecting ABR was unanimous. As board member Richard Shepard, an architect volunteer on the selection committee, said later, "In almost every other interview firms tried to show us their architectural philosophy with slides or illustrations of past projects. ABR showed us overwhelming enthusiasm and told us how they would deal with our problems."

ABR Partner Don Barker says simply, "They turned on to our being turned on."

Museum Director Jennifer Dowley, who sat in on all presentations agrees, "The interview completely wowed the selection committee!"

CASE STUDY #47: SEVEN POINTS MAKE A SUCCESSFUL 15-MINUTE INTERVIEW

Greenville, S.C., is the home of some 22 architectural firms, and because most of the architects are active in community affairs, local public clients have made a practice of interviewing all interested firms before awarding local projects. In practice, this means that an average of 20 firms are invited to 15-minute interviews before each project is commissioned.

One firm that has been successful in winning more than an average share of commissions under these conditions is the Tarleton-Tankersley Architectural Group. Partners H. Harold Tarleton, Jr. and John A. Pinckney, Jr. have developed a seven-point format they feel has greatly improved their batting average.

Recently they had an opportunity to demonstrate the format at an interview for a new $300,000 City of Greenville Neighborhood Center.

Before the interview, the partners visited a similar, completed facility, and talked to the staff to learn the program. The architects also visited and photographed the site of the proposed facility. Then they spent "about an hour and a half" discussing the interview and agreeing on an outline.

"In 15 minutes you don't have time to sneeze," says Tarleton. "We always go to a short interview as a two man team, so while one of us is talking the other has time to watch the committee for reactions. If one of us puts them to sleep, the other one wakes them up."

The outline for the Neighborhood Center presentation:

1. "Jack introduced who we are and passed out business cards. Meanwhile, Harold was setting up the projector."
2. "Jack talked about our present workload. Everybody wants to know how busy we are and how quickly we can get on their job."
3. "Harold discussed our previous related work in the local area."
4. "We addressed general points of architectural concern in the project, such as what we look for in the site. We passed around a brochure of the site pictures. When we talked about the program, we tried to use the very same words they used in their preliminary program. Then we showed highly selective slides tailored to hit their particular problems."
5. "We told them why we wanted the project—because it would be a challenge architecturally, and would be for a public client."
6. "Jack invited them to consult our references."
7. "Harold invited them to visit our office. We noted 'our time is up'; handed each of them a special brochure made up for the project with the committee member's name on the cover; then we got out of there real fast."

Afterward, James H. Simkins, a member of Greenville City Council and Chairman of the Building Committee, explained why The Tarleton-Tankersley Group was selected over 19 other firms to design the Neighborhood Center. "Their interview showed us they had done some homework in preparation, and had been to the site. We don't want firms to get into heavy philosophizing or recitations of jobs they have done. It's most important they they show a feel for the needs of the client and the project we are getting into."

CASE STUDY #49: BREAKING THE ICE WITH A TOAST

The principals of a 30-person architectural firm are alumni of their state architectural school and during a generation of practice have watched in frustration as job after job at the university was commissioned to other practices.

Therefore it was a special occasion when the firm (after diligent bird-dogging) was invited to be interviewed for design of a new music building at the university.

The architects prepared a detailed presentation for the formal interview, but as they developed their strategy they also searched for an appropriate way to express the significance of the occasion.

In due course the time arrived and the architects entered the interview room with some unusual accessories (in addition to their usual slide projector, brochures, etc.). One architect carried a silver tray of stem glasses, one of which he placed in front of each committee member. His partner followed, a white towel over his arm, and filled each glass with sherry.

When all the startled committee members had been served, the architects poured themselves each a glass, and the senior partner smiled and broke the silence:

"Ladies and gentlemen, before beginning our presentation we wanted to take just a moment to toast the significance of this occasion for us, which is the first time in our 21 years of practice that we have been invited to an interview by our alma mater."

They then proceeded with their formal presentation. Shortly afterward they were advised that they had been unanimously selected to design the project.

CASE STUDY #92: A PROJECT MANAGER'S 15-MINUTE INTERVIEW

Last winter a 400-bed general hospoital in New York State shortlisted four firms for master planning and subsequent design work. King & King, a 50-person A-E firm in Syracuse, was one of them.

Initially, the hospital building committee called in each firm for a 10-minute interview. Partner Russ King made King & King's presentation. Following the interview changes occurred in the hospital's board, necessitating a second interview—this time for 15 minutes, with the injunction "We want to see the man that's going to be handling the job—not the partner."

King and project manager David McPharland put together an interview outline. "We had six points," says McPharland. "First, we weren't new to their hospital. Second, our repeat clients were evidence of our ability to deliver high quality service. Third, we gave them two copies of our response to the standard AHA questionnaire. Fourth, we covered my experience in hospital work. Fifth, my experience as it specifically related to their needs and interests. And sixth, a brief outline of the method we use to get projects on the right track from the start."

At the interview McPharland did all the talking, following his outline. "I started out O.K.," he recalls, "but very soon my mouth turned to cotton and I had trouble getting the words out. So I said to them, 'I'm not accustomed to doing this sort of thing, and in fact if you hadn't asked I wouldn't be here. Giving interviews isn't my long suit. But what I can do well is design hospitals.'"

The committee awarded King & King the job, and Dave's been promoted to Director of Operations.

CASE STUDY #93: EAVESDROPPING AT AN INTERVIEW

Last November interviews were held for a feasibility study for a state maximum security institution for the criminally insane in Minnesota. One of the observers in the public interviews was Joanna Baymiller, Public Relations and Marketing Training consultant for Hammel Green and Abrahamson of St. Paul, one of the competitors and joint-ventured with Kaplan McLaughlin of San Francisco.

The following are excerpts from her memorandum to the HGA marketing team.

" . . . The interview took place in a too-small room which presented close-quarters contact between the interviewing firms and the committee (12 people)."

Firm A. "Two very nervous people. No visuals. No slides. Although this firm has had prior involvement with the hospital, they had no additional background for this job. And they allowed their lack of experience to show. They talked about feeling 'comfortable' with the building and with the work, but appeared nervous and uncomfortable throughout their presentation . . ."

Firm B. "X came on as a cross between Attilla the Hun and an encyclopedia salesman . . . a rehearsed, almost rote-appearing style . . . like a routinized giant.

"In addition, X is either very bravado or he mutters. The two younger designers used a large visual spelling out the tasks / time frame . . . They then explained the role each would play . . . It took longer than necessary, but it was helpful to have each person spell out what their role and responsibility would be. X showed slides . . . including buildings specifically designed by the team members present, another good point. However, X spoke as if the firm had already been hired . . . and this was a bit presumptuous . . . They were too wound up, as if they might leave the room and, from the exhaustion of their presentation, collapse like sprung springs."

Firm C. "Y introduced the firm and showed slides. Mentioned . . . energy conservation and that security was 'part of the program.' They showed models of the flexible design process for the hospital.

"Y was nervous, and showed it when Z, who looks and sounds a bit like Frankenstein, made a couple of verbal faux pas ('we don't want to say that, Y grimaced'). He never looked very sure of himself, and seemed a bit embarrassed again when he said 'we're not slick salesmen but we would like to think that we are able architects and professionals.'"

Firm D. (HGA / K-M). "Group took command immediately, reorienting the room to show slides and taking chairs so all could be seated. Bruce began by spelling out roles and responsibilities of the team . . . Mentioned George's background, experience, and expertise. George's opening mentioned how we 'had approached similar projects and would approach this one' . . . Committee attention very keen.

"John spoke with assurance but in a modest way. He referred to the firm as having written a book which was now used as a guide for community mental health centers. Immediately the firm looked like experts. John's comments on 'how we would work with you' caught the attention of the two younger members of the DPW. John's slide presentation referred to specific aspects of the work shown which were pertinent to the St. Peter problem . . .

"Another successful aspect was his emphasis on 'results.' He spoke of buildings which have helped 'contribute to a higher rate of success' in curing / enabling patients to return to society. Also said the firm evaluated their buildings after they were built. Added firm's concern for client satisfaction after the building is a finished product . . . Slides showed several different and dramatic solutions to the design of treatment or correctional facilities. Bruce and John alternated in describing them. Security was mentioned (this turned out to be a major concern of one committee member) . . . The administrator sat listening intently throughout . . ."

John Boerger of K-M, who gave the presentation with Bruce Abrahamson and George Riches of HGA, remembers: "After the 45-minute presentation we asked for questions and there weren't any. We went out shaking in our boots—we either bombed totally or were fantastic, and we didn't know which."

Baymiller's memo concludes: "So why HGA? . . . Presentation was concise, but thorough. Ended on a high note and left the committee literally speechless. Group did not need to say 'we're good, choose us', or 'we really want this job', or 'please be nice to us'. Simply stated an excellent case and left almost no room for choice."

A final note: The chairman later called Bruce Abrahamson and told him that the committee called it "the best presentation they had ever heard."

CASE STUDY #95: THREE STRIKES AND YOU WIN!

Over the last two years TMP Associates, a 60-person A/E office in Bloomfield Hills, Michigan, has evolved a process for preparing for interviews. The approach, says Director of Marketing Martin McElroy, "is something that just happened." But it's so successful that it has become standard practice.

"Three in-house planning meetings precede every interview," explains McElroy. The first is orientation, attended by everyone who will be at the presentation or who knows something about the project. At this meeting they lay out the "message options" ("what is the one thing we want the committee to remember when we leave the room?"). "We might give them priorities—or even decide the one best message—at this meeting, but by the time the options are defined everyone is usually too tired to proceed," noted McElroy.

At the second meeting, the group decides the "best" message and how to hone it. They also review the structure of the interview and identify who will say what.

The third meeting is a rehearsal. The day before, a few people in the firm are assigned as "auditors" who take the role of the client group during the dry-run. Says McElroy, "We look for people who might not be part of our marketing organization, but who could conceivably be members of a selection committee. It might be our president, a project manager, a secretary." The auditors have a day to prepare tough questions and try to throw the interview team off during the rehearsal.

The three meetings are usually held on three successive days. If the schedule is tight, the team may telescope the process into two days, but the three steps are still articulated.

Has it paid off? TMP's interview success rate has been 80 percent.

14

CLOSING THE SALE

"The firm we chose showed they cared about us. The other two firms only cared about getting the job"
Member of a selection committee

What occurs in the period immediately following the interview can be the deciding factor in the business development process. At this stage, however, the initiative is no longer entirely in the professional's control.

With one exception, all the cards are now in the prospect's hand. The client will decide whether to play another round and, if so, what the rules will be. The professional must be prepared to play whatever game the client deals.

One last card the professional can play immediately after the interview involves writing a summary letter making a final, strong pitch for the job.

The purpose of such a letter is three-fold. First, it serves as a courteous "thank you" for the interview. Second, it allows the professional to restate briefly the particular strengths which qualify the firm for the job. Lastly, and possibly most important, it is a final opportunity for the professional to demonstrate genuine interest in the assignment.

Where appropriate, this letter may serve to save the project that is slipping from the professional's grasp. This can occur when, after an interview, the professional suddenly realizes that a critical point may not have been covered.

An example of this occurred to a joint venture of architects and engineers who were applying for a federal planning assignment. Five professionals represented the combine at the interview. Immediately afterward they held a postmortem (always a wise step) and realized they had completely overlooked discussing their strengths in accurate estimating and cost control—a critical requirement on this particular project. A summary letter was therefore composed which led off saying: "We feel we may not have adequately stressed our experience and qualifications in budget control . . ." and continued with a page and a half of supporting examples. In due course, the project was awarded to the joint venture. While there is no proof that the postinterview letter tipped the scales, the federal agency's press release announcing the selection said the joint venture had been chosen because of their "strong record in cost control."

Another point worth stressing in the postinterview letter, if appropriate, is an invitation to the prospective client to visit the firm's offices and to tour projects it has designed.

An alternate to the summary letter, used with considerable effectiveness by some firms, is to keep minutes of the interview and then mail them in memorandum form to the committee. Such minutes note each question asked and record in detail the firm's answer.

The summary letter, or minutes, should always be addressed to the chairman or chief administrator of the selection committee. Individual strategy will determine whether it may be wise also to send copies to all members of the committee. Although, as discussed previously, this volume does not attempt to suggest reliance on personal contacts with influential persons, it is worth mentioning that if such people are involved it may also be appropriate to send them copies of the summary letter, so they will be equipped with the same ammunition to use on the firm's behalf.

VISITS TO PROJECTS

It can be hoped that the client will elect to visit completed projects designed by the firm. There is probably no better way to judge an architect or engineer's work. But few selection committees take the trouble, although the number does appear to be increasing in recent years. It is for this reason the professional who has good work to show should use every opportunity to urge the selection committee to see it.

Selection committees which do visit completed work usually are of two schools on procedure: Some, in an apparent effort to assure they learn the "truth," believe the visit should be made without the professional present; others welcome having the professional serve as their host and escort.

In the former case, little can be done because such committees do not normally even tell the professional that they will make the visit. These committees look on the inspection as a sort of intelligence investigation, and completely miss the more important point that such visits are a very valuable test of human chemistry between themselves and the professional they are considering.

It is for the latter reason that the professional should always try to escort the prospective client personally through the firm's work.

When the opportunity is presented, a few simple rules can be applied:

a. Project visits should be chiefly visual, rather than technical, experiences.

b. The length of the visit, therefore, will be determined by how long it takes to walk through the project and see its significant features.

c. For laypeople, a tour of 30 minutes to an hour is long enough. If technicians with special interests are in the group, provide them with a separate escort so they can concentrate on their areas of interest.

d. If the project is of the same type being considered by the prospective clients, they may welcome the opportunity to discuss programatic information with the users. Therefore, if the owner / occupant offers an escort, try to have it be a representative of management, not someone from the building maintenance or operation staff.

e. Do not ever ask the owner to "lead" the tour. What the proud owner wants to show will often be less than pertinent and will take too long.

f. Above all, the professional should present the project to the inspecting prospects in the same "problem solving" approach as discussed in the previous chapter: State at the front door the program you were asked to solve, and then explain the functional solution as you go through the facility. There is little to be gained by extensive discussion of details, systems, methods, or materials whose choice was not essential to the solution. The prospect's questions will determine any interest in these areas.

g. Whatever is planned, dry-run it in advance with the project owner, or send an advance scout an hour ahead of the prospects. The best plans laid by phone or letter have an uncanny way of being misinterpreted, resulting in hour-long waits for a key to unlock the mechanical room.

In summary, the important objective of a building tour is not to bore the client to death. Arrange the visit to move rapidly, and let the client slow it down at points of unusual interest.

The other guideline, which should hardly need to be mentioned, involves application of simple good manners. The professional should assist the touring committee with transportation, reservations, and any other appropriate details of logistics which may be helpful to them. If several stops are planned on a half-day or day-long tour, give each visitor a typewritten agenda in advance. Small as they are, these details contribute significantly to the chemistry of client-professional interaction. Ask anyone who has toured the work of three or four professionals—the one who makes the visit the most pleasant experience will very likely be the longest remembered.

Whenever the logistics involve out-of-town travel, overnight lodging, or major meals, there is always the sticky question of who picks up the check. There can be no hard and fast rules in this area, because there is a considerable latitude in the ethical views of different prospective clients, and there is also a great range in the ability of different salespeople to carry this sort of thing off. Wherever possible, however, the very best approach is to arrange those things you want to provide gratis so that there is no opportunity for the question of payment to arise. This means use of company or chartered airplanes or limousines for transportation, meals at private homes or clubs, and entertainment by prepurchased tickets to theatre, music, or sports events rather than at nightclubs. If the major expenses can be handled in this way, then there should be no hesitation about letting the prospect pay for lodging, incidental meals, and other direct expenses.

OFFICE VISITS

If the committee agrees to visit the professional firm's offices, the best rule is to be relaxed. A certain amount of advance tidying is probably in order, considering the nature of architects' and engineers' businesses, but if the office looks like a stage set the visitor will spot it as phony in a minute. The staff should continue working on whatever they are doing. The visitors will be very interested in seeing and learning about the work in progress. Be sure the models or appropriate schematic drawings of interesting projects are out in the open in the drafting room. Then it is easy to pause during the tour and explain one or two particularly interesting projects. It is not necessary—and can be awkward—to ask the nearest job captain or draftsperson to explain "their" project, but by all means introduce them by name and qualifications (e.g., "Jim has been with us for 10 years and knows more about codes than anyone in the city" or "Jane joined us after winning a traveling fellowship to study solar design in Egypt" etc.).

The tour of the physical office should be supplemented by a further presentation around a conference table of the firm's organizational structure and the methods by which the project under consideration will be handled.

It is also proper to encourage discussion of how the client will organize to staff the job. The more the client can be made to visualize what it would be like working with the professional, the more comfortable the client will feel about selecting the firm.

And small firms should not be shy about the size of their office—even if they have only one or two staff members. If the firm is qualified at all to do the work it is seeking, the client will not be "turned off" by a modest organization.

Throughout the period when the client may be touring completed projects or visiting the professional's office, the hard sell should be avoided. If the client has come this far, it can be assumed that there is genuine interest in the firm being considered. It is no longer necessary to sell the basic merits of the firm. Now is the time for human chemistry to bring about permanent chemical change. This always happens best in a natural environment.

CONTRACT PROPOSALS

The practice of clients asking for formal proposals or draft contracts prior to final selection is increasing, and such requests represent a particularly treacherous aspect of closing a sale. At this point, generally, one has survived the qualifications screening and has passed the "chemistry" test: The client is prepared to work with you, but wants to know the hard terms and conditions before making a final commitment.

Some firms simply use standard contract forms as their response when asked for a formal submittal at this stage. While it may be appropriate to attach a copy of the form of contract to be used, it is rare that the contract alone will suffice. Contracts are not selling documents. The mere length of a contract is often intimidating to the prospective client, and the language tends to be less than stimulating. More serious, from a sales standpoint, is that the conventional contract forms tend to minimize the professional's services, in a legally legitimate effort to set boundaries on the scope of what the professional can be obligated to do. The AIA "Standard Form of Agreement Between the Owner and the Architect" is a classic example.

The great failure of the trend to more legalistic approaches to professional practice in recent years is that professionals have become preoccupied with what they will *not* do at a time when society and the building industry are trying to expand what they *will* do.

Consider, for example, the supplemental architectural services such as programming, interior design, landscape design, etc. The AIA's B331 Owner-Architect Agreement takes great care to identify these as "additional services" which the architect will perform *only* if specifically authorized and if compensated for on a separate basis. This is a poor way to sell anything—and if there be any doubt just consider the tremendous recent growth of independent organizations who specialize in space planning, interior design, land planning, etc. Bluntly, the architects who preach "comprehensive services" and the return of the "master builder" will never make it with AIA Document B331.

Many architects and engineers have already gotten this message and have scrapped standard contract forms in favor of more sales-oriented documents. The basic requirement of such documents is that they should begin with positive statements of what will be performed. One good approach, if one's lawyer approves, is to write a letter agreement to which is attached a boilerplate statement of "Terms and Conditions." Another is to draft a custom contract document for your firm and submit a completely individual contract to each client when requested.

Much of the interest in preselection proposals, of course, is in the price (see below). But don't miss the opportunity to make the proposed contract itself a selling document. If it takes only one straw to tip the scales in the final selection, be sure this one is a straw in your favor.

SELLING THE FEE

Ever since the revolution in professional ethics brought about an end to collective fee schedules (Chapter 2), more and more design professionals are being required to submit price proposals prior to selection. The trend has caused turmoil and trauma in some professional offices, while others have just taken the whole thing in stride. The difference is important to understand if one is to be comfortable and successful at marketing work where price is a consideration.

At the outset it must be emphasized that being low-bidder is not analogous to selling a fee. If one has done the best job of selling one's charges, and then also comes out low among the competitors, so much the better. But the great majority of professional selections do not go to the lowest price. Clients are quite aware, as architects and engineers have maintained for years, that quality professional services cannot be obtained on price alone. If the client wants the service of one firm enough, and its charges are presented in a believable and persuasive manner, it will get the job even when it quotes the highest price.

That is the chief point to remember when selling a fee. How to do this can take many dimensions depending on the circumstances.

Quite often price quotations are requested before enough is known to specifically quantify the work required. In such cases, one of the best approaches is to quote a range, such as:

These services can be performed at a maximum fee of $_____ to $_____ depending on your clarification of the following questions. . . .

or

We have designed past projects of this type at a fee between _____ % and _____ % of construction cost, with the range depending on how much of the effort will be done by your team and how much will be entirely our responsibility.

When answers of this type won't work—and sophisticated clients are getting more particular about what they ask for—there are a range of devices that can be used to make your fee quotation stand out in a crowd. Among these are:

Publish a company fee schedule giving personnel rates and defining the basis of charges, thereby giving the client the feeling that whatever lump sum you quote is based on charges comparable to those paid by other clients.

Quoting unit fees per square foot of building or per gallon of treatment capacity, etc., thereby giving the client with an incomplete program a way to assume some control over the final cost.

Etc.

Another avenue to selling the fee can be to ask the prospect what was paid for previous professional design services the client commissioned, and to inquire how the client believes the job at hand compares to that project.

These are just a few of the approaches that can be used to sell adequate compensation for your work. Negotiating is a skill of its own, which can be learned. If one has doubts about one's negotiating ability one should go to specific seminars on negotiation and read some books on the subject. Professionals who have taken the trouble to learn the skills of the negotiating game report excellent subsequent results when it comes time to sell a fee.

Above all, don't cut your price just to win work. Very few of the construction contractors who play that game enjoy a reputation that approaches being "professional." If you are worth what you must charge to be fairly compensated for what you do, quote fairly, market the value of your services as hard as you can, and you will win plenty of work. The design firms that have taken the new era of price quotation in stride are those that offer a top quality service and know how to sell it. When that kind of marketing is being done, price is a minor consideration in the final selection.

WATCHFUL WAITING

As the last hand in the selling process is played, the client for the first time will be under more pressure than the professionals seeking the commission. The client must make a decision—and this often means much more than selecting the architect or engineer.

Any number of factors have an uncanny way of disrupting the client's timetable at the moment of truth. The professional firms may have been promised a decision by a certain date, but the decision means spending money. Perhaps the client must now return to a board of directors for appropriations. Internal politics may also become active within the client organization, particularly if several professionals are playing "who-do-I-know."

Even if one gets a phone call saying that your firm is on top of the list and will be presented to the Board of Directors next month, it is not wise yet to turn down other incoming work or to order that new sports car. There are always a few jobs which go all the way through the selection cycle, and then never get off the ground.

Do not be dismayed if a job which seemed so urgent suddenly slips into limbo for weeks or months on end.

It is quite proper during the "limbo" period to call the client now and then and inquire if there have been any new developments. In doing so, always ask the client for an estimate of the new timetable, so you will have an excuse to call again if the next stipulated date slips by. But don't pester— the client who cannot keep to a promised timetable is probably embarrassed enough. Meanwhile, if something happens in your organization which may add to your case—such as receipt of an award, or hiring of an important extra talent—it is good form to drop a note to keep the prospective client informed.

All of which adds two more axioms to the list of business development rules: Don't ever plan on a new job until it is actually underway; and don't ever count a job out until it is officially decided in favor of someone else.

from: **the Coxe Letter**

CASE STUDY #18: INCENTIVE FEE WINS A JOB

In 1973, the Hillsborough County (Fla.) School Board urgently needed to build an elementary school on a tight budget. Tampa architects Rowe Holmes Associates applied for consideration. The firm had designed another school for the district two years earlier which had yet to start due to site problems. In the interim, H. Dean Rowe, AIA, kept in touch with the school people, and felt he knew enough about the client and program to offer the District an unusual proposal.

The board wanted an economical solution and occupancy within a year. The budget was just under $2 million ($27/SF) including furnishings, equipment, landscaping, and site work. The District's standard fee was 6% of construction costs.

"We told the School District we could do the job," recalls Rowe, "but we said we would be cutting our own throats if we saved them money." Therefore, Rowe Holmes offered to do the project under an incentive contract that provided:

- A 6% basic fee for all services;
- A bonus fee of 10% of all savings under the School District's budget;
- A penalty clause to reduce their fee by 10% of any overage.

The job was awarded to them, on this basis, on Sept. 5, 1973. Fast-tracked under seven different bid packages, the project was scheduled to open in early October, 1974. Final costs (except fee) were about $23/SF, or $250,000 under the budget.

The architects are pleased with the job. "Had we not had the incentive we would have lost our shirts," says Rowe, "but under the contract we were able to spend more to do it in less time. Given another project with a defined scope and realistic budget, we would be happy to use the same arrangement again."

The Hillsborough County School District was apparently also delighted. They budgeted three more elementary schools, and were talking to Rowe Holmes about repeating the firm's design and process on at least two of them.

CASE STUDY #42: ITEMIZED PRICING PLUS INCENTIVE BECOMES PROMOTIONAL TOOL

The first job that Newark, Ohio architects Harden & Reid went after following their introduction to the new AIA Compensation Management system was the remodeling of the Coshocton County (Ohio) offices.

When partner John W. Reid, AIA, heard the county was planning to do something, he visited the commissioners to put the firm's hat in the ring, and was later one of a number of architects invited back for interviews. Subsequently, Harden & Reid and two other architects were asked for proposals.

Reid returned to the client and, following the AIA Guidelines, "gathered enough information to make some assumptions on how the job would happen."

The firm's proposal itemized each phase of the work by the estimated number of hours required, and multiplied the time by a flat (average) billing rate of about $23 per hour. In addition, the number of trips to the job, the quantity of prints to be supplied, and a contingency were listed to reach a total. The firm then proposed to do the project on an hourly basis with that maximum amount, with any savings going to the owner.

Harden & Reid was awarded the job. "We asked them later about our proposal and they said they had never seen anything like it before," says Reid. Compared to percentage fee quotations, he says, their proposal "right away gave the County a figure they could budget plus an added incentive."

"It's a hell of a promotional tool," says Reid. "With a lot of clients in small areas like this it comes down to price. When you spend the time with the owner to gather information for such a proposal they are usually very interested. Then when you give him the incentive to get the job done quicker he sees the possibility to save money on this fee he doesn't want to pay anyway."

"We've been using the approach on everything we've gone after since the Coshocton job," says Reid. "We've found it works." Their 5-person firm has just been awarded a similar job in a neighboring county and is looking forward to a "better winter" than anytime in the three years since the partnership was formed.

CASE STUDY #68: TAILORED PROPOSAL BEATS LOW FEE

When the Fullerton (CA) City Council sought a consultant for the design work for central business district improvement, nine firms, including Bissell-August and Associates of Newport Beach, responded with proposals.

"We always go right to the redevelopment agency or committee before writing a proposal," says George Bissell. "It often gives us some insight on the approach to take. It also shows that you are interested in their project."

On the visit to the Fullerton Redevelopment Agency, Bissell-August told the client that the RFP was "backwards"—that Phase 1, the development of three alternative design themes, and Phase 2, a detailed inventory of existing facilities, should be reversed. In its proposal the firm reiterated its conviction.

Although another firm submitted a lower fee, according to Council spokesman Paul Berlant, its proposal "did not express the same understanding of our particular needs as well as the Bissell-August proposal." Berlant continued, "To select solely on fee in this case would be a mistake. The proposed work outline and experience of each firm are probably the most important factors to consider."

The City Council voted unanimously in favor of Bissell-August.

CASE STUDY #71: BRICK IS MEMORABLE LEAVING PIECE

Last fall officials were selecting an architect for the new Frederick County (Md.) Courthouse. The selection committee included state and county officials, two judges, the state's Senators, and several members of the historical commission.

The courthouse site in Hagerstown is in an historically-certified district, and the selection committee wanted an architect who would design a structure compatible with and sensitive to the mixed historical and commerical elements of the site.

Ten firms were shortlisted, including a joint-venture team of Bushey and Burrey of Hagerstown and The VVKR Partnership of Alexandria, Va.

After showing slides (addressing the client's historical concerns and not emphasizing the team's substantial courthouse experience), Jack Burrey presented to the committee chairman a Bicentennial brick produced by a local 100-year old brickworks, saying, "We'd like to donate the first brick to this project."

"The project needed to have the flavor of the past," says Burrey. "We wanted the committee to believe that this wasn't just another pitch, and that we were really interested in the project. The brick was something to remember us by." After a second interview, Bushey-Burrey / VVKR was awarded the commission.

CASE STUDY #83: COMPREHENSIVE SERVICE WINS JOB AT HIGHEST FEE

The School District of Union, Missouri recently invited three architectural firms with known planning capabilities to interview for a study of the District's long-term and short-term needs. Superintendent Chris Straub chose the firms because one had done "a super job" on a recent remodeling project for the district; another had birddogged the district and demonstrated its interest; and a third—William B. Ittner of St. Louis—had conducted a similar planning study for the local community college.

At Straub's request, each firm spent a day with him visiting all the buildings and discussing what the School Board thought was needed.

Afterward, Ittner President Bob Little recalled his reaction: "Most of the people on the Board didn't know what their needs were. We helped the superintendent understand what was really needed." He proposed to do a demographic survey, a space needs inventory, and studies of grade structuring, handicapped provisions, and food services for a basic fee of $5500, plus an energy analysis of five buildings for $2500 additional. Another competing firm offered to do all the physical planning work for $5600 and suggested an outside demographer for that survey at a separate $2000 to $2500.

The third firm offered to do the planning work and a brief energy analysis, but without a demographic study, for $3900.

The local newspaper, the Franklin County Tribune, reported the deciding discussions at the next School Board meeting:

"In discussions following the architects' visits, the board felt that energy and demographic studies were essential items in the survey."

"The other firms (excepting Ittner) screened by the board . . . offered to do little or no energy or demographic studies."

The paper quoted Straub, "It seemed to me that (the low price firm) was using the study as a way to get his foot in the door."

Finally, "(Board Member Bill) Walker made a motion to accept Ittner, Emmett Fink seconded the motion, and the board voted five to one in favor of the motion."

CASE STUDY #86: LAUNDRY LIST PROPOSAL PLUS ENTHUSIASM LAND A JOB

KZF, a 120-person A-E-P firm in Cincinnati, was one of 7 firms invited to submit proposals for a downtown revitalization study for the city of Trotwood, Ohio. The first decision, recalls Curt Velsor, landscape architect and project manager with KZF, was whether or not to go after it. "The planner in another city where we were working referred us to Trotwood's city planner. Their budget for the creation of alternative concept plans for the Old Town area was originally $2500. The project was much smaller than those we generally go after, but because it was a referral by another client, we didn't want to decline."

A member of KZF marketing and sales staff contacted the city's planner, Frank Raeon, and offered to send information on the firm. Shortly afterward Velsor and the marketing person visited the site, with Raeon, to find out what they wanted.

The subsequent proposal "was more than they asked for," Velsor observes. It included an inventory of the physical condition of Old Town and the market potential of the city; development of alternative concept plans; and an estimate of implementation times and costs. It spelled out elements of the work and the consultants' estimated fee for each, totaling about $6500.

KZF was one of four firms invited for an interview. Velsor set up presentation boards showing KZF's work (they had been prepared for use at a recreation convention a few months prior) and talked from them. "I gave them a kind of status report on each project," he says. "I pointed out instances where KZF made mistakes, and where the client may have done something that cost him more money than it needed to. The focus was on learning from our experiences."

Velsor then summarized the firm's proposal, with copies at hand for the committee's reference (they had all received—and read—it beforehand). "I didn't try to sell the whole proposal," he observes. "I emphasized that clients are their own best resource and encouraged them to identify the elements in the proposal that they could do themselves, and to use a consultant only for the portions they felt they weren't equipped to handle. The message was, 'Don't buy the whole package; buy what you need.'"

He then walked them through the process of planning and working with the client. An open discussion of the city's project ensued. The committee let it run 30 minutes over the allotted 45 minutes.

A few days later Raeon notified Velsor that KZF had the job. (The client eventually selected services totaling $4200.)

Velsor then asked if Raeon would write him a letter about the committee's reactions and why KZF got the job.

Raeon's letter divulged that KZF had been ranked #3 before the interview. What won them the job, he said, was Velsor's demonstrated enthusiasm for the projects he showed (the committee felt as though it was the kind of status report they'd get on their project); the fact that he established a dialogue; and the way in which he presented what the city should be looking for in a consultant and their options for implementing the work.

CASE STUDY #41: CHARGING ON WORTH VS. COST

"The practice of telling a client you will charge a multiple of your cost is ridiculous," says Atlanta architect Henry D. Norris. "It's none of my client's business what my costs are!"

Architect Norris, who has been in practice since 1950 doing mostly developer housing and single family residences, has 15 years of successful experience with flat rate fees to back up his strong convictions.

"I tried every way of charging when I was first in practice," Norris says. "Then one time I was doing work for a prefab manufacturer and the president came to be a close personal friend. One day I asked him what he thought about my fees. His answer was: 'Anyone who doesn't charge $250 a day isn't worth the money.'"

Since then Norris has charged flat rate fees. His rates have risen with inflation to $60 per hour for the principal; $45 for registered architects; and $30 for draftsmen. "Everytime we raise them we seem to get more business," he says. "We have met virtually no opposition to these prices. We've had a few chiseling type developers who think it's sort of high, but the quality clients who are the kind you want to work for don't question it at all."

In making proposals, Norris says, "Everybody wants some idea where they are going." Therefore, his agreements state: "Based on past experience our charges will probably not be less than _____ % or more than _____ %, but we do not guarantee these ranges." Norris then tells the client what will vary it, and says, "When we outline the variables we don't have anyone question the range."

"If you make a commitment to work this way," says Norris, "you have to rule out government work. But it doesn't bother me at all that they won't give me fees on this basis because I don't want the work. I did a public housing project; I lost money; and I'd never do it again."

Otherwise, says Norris, "We'll do anything, even a screen porch. Working on an hourly basis there is no job that is too small. You make a profit even on a half-hour consultation."

Norris isn't bothered that his pricing practices are different from other architects. "No one has ever questioned it. My lawyer doesn't charge a multiple of cost. Other businesses don't do it. People have told me time and again when we were through that they thought our service was well worth the money."

15

WIN, LOSE, OR DRAW

*"You can do an awful lot of business development
without developing any business."*

Discouraged marketer

Sooner or later, the day of decision will arrive and the professional will get the word—sometimes by telephone if one is selected; usually by letter if one is not.

The winner's work has just begun. After all that has been put into getting the job, let us hope that doing it will not be an anticlimax. If it does go well, the architect or engineer will have written their own textbook on "How to Get the Second Job—for the great majority of every professional's commissions come as repeat assignments or by referral from satisfied clients.

The loser should not be totally dismayed—the law of averages were against winning from the beginning, but they shift toward one's favor each time one comes in second.

The business development game should never be played with a single fish-hook hoping for a lucky strike. The professional fisherman sets out a boatful of lines and tries to keep bait on every hook. Then, when the fish begin to bite the catch is counted not by ones and twos, but by the pound. If the bait should slip off any one hook, the angler will probably never know the difference.

It is no disgrace to be second or third choice from among two or three dozen initial candidates. There are far too many variables in the process and its chemistry for any single firm to be the "right" one in all cases. A firm which consistently reaches the interview stage should expect to receive at least a relative share of commissions.

The firm that really needs to worry is the one which is never invited to an interview or, if interviewed, always strikes out. In such cases, the firm is either seeking work completely over its head, or it is doing an inferior job of presenting itself.

For these reasons, the losing firm should keep its file open until an attempt has been made to find out why the decision went the way it did. Often this can be the most important step of all in the business development process.

If tactfully done, it is not half so difficult as it might seem to find out why one has struck out. A phone call or a visit to the prospect will reveal useful information in at least half the cases. But don't ask "What did I do wrong?"—that is perhaps an impossible question to answer. Ask instead: "Could you tell me what were the principal factors which influenced the final decision?"

With that as openers, many clients will be happy to be frank. Often, the reason will turn out to be factors beyond the control of the losing firm. But if the firm lost because of a midjudgment in presenting itself, invaluable lessons can be learned.

Another very valuable question to ask in debriefing is: "What was the nature of the presentation (or services offered) by the selected firm?" If you can find out what your competitors do right when winning jobs, that information is far more valuable than being told what you did "wrong." This approach is especially worth remembering if you seek the kind of debriefing many federal government agencies now offer losers as a matter of policy. In such cases the debriefing officer will go over your evaluation sheet and show you where you scored low. But the most valuable information to get, if you ask, is what the winner did / said / presented / proposed to score higher.

An example of the value of "debriefing" a lost commission occurred following selection of the architect / engineer for a very large urban medical center project. Two out-of-town firms were in the final running. One had great experience in the building type, plus other very strong qualifications, and they seemed to have it in the bag. But the choice went to the second firm, in spite of what appeared on paper as lesser qualifications.

The reason, learned by the losers because they took the trouble to ask: The winner had spent several hours before the presentation walking around the neighborhood in which the project was to be located. During the interview this professional impressed the selection committee with both his grasp of the challenge of the project and his interest in it. Axiom Number 101 was thereby added to the loser's rulebook: Always inspect the site before an interview.

Debriefing can be useful to the winner as well as the loser. Successful firms will often be quite surprised by what they learn when they ask their clients what specific factors led to their selection.

It is often said that selling is a game with no second prizes. The author believes there is a corollary that there are no real losers in a sales contest among professional firms. Those who are not initially selected still come out ahead in many ways. They have gained sales experience which will certainly sharpen their approach on the next attempt.

And, often more important, the losing firm will have exposed its qualifications to a significant group of people who may have something to say about selection of professionals for other jobs, at other times, and in other places.

The real lesson of the business development process is that professionals can and should be more than merely responsive to potential new business. It is not all a matter of "tricks of the trade" or "who you know." Those who are aggressive to the opportunities to pursue prospective clients can do much to increase the quality, quantity, and interest of the work they do.

In this very real sense, once a professional decides to conduct a formal business development program, it does not greatly matter whether he or she is selected for any one particular job. Win, lose, or draw, the committed professional will find more than enough incentive to try again.

from: **the Coxe Letter**

CASE STUDY #52: A BACKWARD LOOK PAYS DIVIDENDS

Recently the Syracuse architectural firm of King & King was one of the finalists interviewed by the U.S. Postal Service for a major mail handling facility in upper New York State. But the commission was awarded to another firm.

"We felt we had done an especially good presentation," recalls Jeanne Murphy, King & King's assistant director of marketing. To find out what had gone wrong, she decided to take advantage of the new policy of most government agencies that allows architects to receive a debriefing after the selection process is complete.

Ms. Murphy went to Washington and visited George Williams, Manager of the Postal Service's Building Branch.

"I learned that out of five finalists, we came in fifth," she reports. "Mr. Williams got out the score sheet evaluating our interview and went down it with me, line by line.

"I found out that they thought the delivery of our presentation was very good, but we weren't as heavy in industrial experience as we thought we were; we had proposed a flow chart that was misunderstood by the postal people; and they felt we were evasive about the personnel we would assign to the project.

"We had also been under the assumption they knew a lot about us from our Forms 254 and 255, so we touched only briefly on our organization. I found out from Williams the interviewers don't necessarily remember anything from your 255, and we should have told them all that over again during the presentation. For the same reason, we didn't show them examples of our work at the interview, and we got marked down for that."

Ms. Murphy reports that Williams was very approachable and helpful. "He will tell you anything you want to know," she says.

The debriefing has already paid off for King & King. In late May the firm was one of nine architects interviewed by the Sisters of Charity Hospital in Buffalo. Ms. Murphy reports they applied a number of the lessons from the Postal Service critique when preparing their presentation. King & King was awarded the job and was told afterwards "there was no question" about which architect made the best presentation.

III

Tools of the
Trade

In planning for new business development, the role of the implements of selling should be clearly understood. Brochures, photographs, publicity releases, slide shows, speeches, mailings, advertising, and the like are all instruments of communication, and they can contribute significantly to the selling process when effectively used. But having a brochure is not the same thing as having a marketing program, and issuing a publicity release does not mean one has public relations.

The following chapters endeavor to describe the function of the major tools of communication and how they can be applied most fruitfully to the professional's marketing program. What follows is not, however, intended to be a how-to-do-it guide to teach architects or engineers how to become journalists, publicists, or graphic artists. These are separate disciplines which require talent and training of their own and specialists in these fields are available to provide such services. The architect or engineer who thinks he or she has time to learn these skills would be better advised to devote that time to market research and bird-dogging.

16

THE PRESENTATION SYSTEM

"We think if we set a good standard of professional-ism, then the clients will respond."
                              Engineer William R. Ratliffe

Architecture and engineering are, in themselves, acts of communication. Does it follow, then, that the professional who is adept at communicating designs to clients may also be an effective salesperson to prospective clients?

You bet it does!

It there is one common talent which marks virtually all professionals who are both respected designers and successful practitioners, it is the ability to present their work in a stimulating and persuasive manner. And when one looks behind the effective presentations these offices produce, it becomes apparent that the capability is no accident. Almost without exception these offices have a communication *system* built into the design process. This system is organized so that a presentation geared for the client is the automatic output at each milestone as the design evolves through schematics, preliminaries, etc.

Having such a system is the essential point of this chapter, for it is axiomatic that if a professional has readily available good presentations of all the firm's past and current work, there will never be a shortage of effective material to communicate with future prospects.

Making such a system work is not difficult. It is important, however, that the output of the system be aimed at the lay client. Some architects and many engineers, who have no difficulty communicating with their peers, have a mental block when they try to explain their work to a nonprofessional audience. Perhaps this is an unintended by-product of the way professional schools teach their students to present to design juries. Unfortunately, what communicates with a jury has little to do with what will communicate with a layperson.

On the other hand, the solution is not the "one rendering" presentation which packages months or years of professional effort into a single pretty picture. Communicating with the lay client does not mean to oversimplify, abbreviate, or talk down. Rather, it is a matter of presenting the full technical / professional message in a language the layperson can readily understand. Developing this vocabulary is the essential element of all successful presentation systems.

How does one talk architecture and engineering in lay language? There are numerous effective variations, but the basic vocabulary requires only three things:

Three-dimensional Graphics

A Written Statement of the Solution

A Problem Solving Sequence

Each of these deserves some separate explanation.

GRAPHICS

Drawings are, of course, the basic medium of the business. But there is a considerable difference between those graphics which are technically meaningful and those which a layperson can readily grasp. The practitioner can read a complicated floor plan faster than this paragraph. But the client may become totally disoriented when looking at the typical preliminary plan. The client is not accustomed to scale; confuses outdoors and indoors;

doesn't know that a thick line is a wall and a thin line is a window. The reason is that the layperson generally perceives buildings in three dimensions, and does not instinctively translate two dimensional material into an understandable vocabulary.

The simple answer is to communicate with the layperson in three dimensions. Obviously, however, this is much easier to say than to do. Renderings, one point perspective plans and sections, detailed models, mock-ups, etc, are time consuming and costly to do, provided one can find drafting staff today who are capable of making such documents.

A more practical solution lies in the difference between what the layperson wants to know and what the professional has to tell.

Block models, diagrammatic plans, and simple isometrics will usually tell everything the lay client *wants to know* about what the professional has designed.

Many architects are properly offended by the kind of crude diagrammatics which are used to explain their buildings in newspapers or fund-raising brochures. But this is what the layperson grasps most readily. The details—the things that convert a "scheme" into "architecture"—contribute little to the layperson's *understaning* of the solution. The details may lead to more *appreciation* of the work, but first it must be understood.

Communication graphics must begin by aiming for simple understanding. This does not mean that they need be crude. A very effective client presentation can often be made with only 10% of the effort that one might put into a sophisticated, detailed presentation for a professional audience or jury.

Some of the basic graphic techniques of a simple presentation system include:

Site models. Many offices build a small scale site model as soon as they get a commission—before any schemes are drawn. Block studies made on such models—even at 50 or 100 scale—are both meaningful and impressive to the client. And they photograph extremely well. Such a model, refined regularly as the design moves along, can usually serve the life of the project, not only for client presentations but also for client publicity, display, and (via photographs) for the professional's own sales brochure.

Humanized plans. Two simple elements will usually convert the most complicated plan or section into a drawing that is reasonably intelligible to the layperson. The first is addition of tones or rendered landscaping to separate outdoors from indoors. The second is to furnish interior spaces with scale indications of furniture. The layperson who might be completely disoriented by a two-dimensional plan can instantly recognize and relate to a tree or chair.

Even this is often unnecessary during early design phases. At the schematic stage, blocks of tone or color can be used to indicate different functions, thereby obviating the need for individual layouts. Such plans make excellent color slides, and often succeed in getting across the big ideas so effectively that it never becomes necessary to make presentation plans of the final layouts.

Perhaps the best test of communication graphics is what they mean to someone who has no direct interest in the project, such as a prospective client. If a detailed technical explanation is required each time a drawing is shown, it is probably much more complicated and technically complete than it needs to be. And engineers who think they are an exception to the above are making a grave mistake. Engineers so often work at the project level with other engineers that they tend to forget the importance of communicating with laypeople like mayors, city council, company presidents, etc. But unless the engineer goes out of the way to communicate in their language, the firm may miss the opportunity to be remembered and, consequently, hired a second time.

A final, essential element of the graphic system is a means to record and reproduce the output. Every model and drawing should be photographed or photostated *before* it is shown to the client. This must be standard procedure regardless of circumstances. Original presentation materials have an uncanny way of being loaned, left behind, damaged, or carefully stored away where they are never seen again. And the cost of retrieving or replacing one or two pieces at a later date is much greater than the original cost of photographing the whole set. In addition, photographs are often a far better medium than the originals to leave behind with the original client, because photos can be passed around among the client's audiences, released as publicity, and reproduced in promotion or fund raising brochures.

WRITTEN PROJECT STATEMENTS

The second essential element of a communication system is a *written* summary of what the design is all about.

A classic illustration of why this is important is the following quotation from a brochure issued by Sarah Lawrence College to raise funds for some major new buildings designed by Canadian Architect John Andrews.

"Mr. Andrews then presented his concepts of the complex which will house the new library, art studios, science laboratories, and faculty offices. His explanation was accompanied by slides showing his ideas in graphic form. The largely visual nature of Mr. Andrews' presentation makes reproduction of his remarks in any kind of narrative practically impossible. We have been able to string together a few quotations, however, which give the flavor and direction of his approach. . . .''

When one considers the type of people to whom Sarah Lawrence sends its fund-raising appeals, the magnitude of this gap in Mr. Andrews' communication system becomes doubly clear.

This is not an isolated example and its occurrence is totally unnecessary.

Most architects (including, apparently, Mr. Andrews) and some engineers have a real talent for describing their work in colorful and exciting terms—verbally. As a practical matter, however, the audience the architect or engineer reaches in person is only a fraction of the client's audience. Unless the professional arms the client with written material to accompany the drawings, there is risk of being misunderstood by and/or losing the larger audience.

Trade magazine editors report they routinely fend off invitations to visit projects when the excuse is "you can't possible understand it unless you come and see for yourself." Obviously, if the design cannot be understood from photos, drawings, and a written description sent through the mail, how would it qualify for publication?

The written material that should be a part of the communication system need not be elaborate or lengthy. One or two typewritten pages can be enough. And the content may be very general—the sort of thing that architects use in describing their work for honor award submissions is entirely adequate. The typical outline should encompass a one-paragraph explanation of each of the following:

Program. What was the problem the professional was asked to solve?

Site. A physical description is sufficient, plus unique constraints, if any.

Design solution. What did the professional really do that was significant?

Structure, materials, and systems. Briefly, to the extent they are significant.

Construction. How and when will it be built.

In terms of the professional's public relations, the most important section of this outline is the philosophy of the design solution. This is where the essence of the professional's creative role becomes clear and distinct from that of the technician. The length of this statement is not important. Two or three paragraphs are often enough. But if they are clear and colorfully stated the professional will be quoted over and over again as the project moves along.

The time to prepare such a description is *before* the initial presentation of the concept to the client. Then the client will be armed with communication materials when still excited from the professional's verbal / visual presentation. Having the written statement at this time will often generate other effective uses for it. It can be combined with photographs of the graphics in a loose-leaf binder and copies given to key members of the client's staff and board. It can be quoted in publicity. And should the client decide to publish a financing, leasing, or public relations brochure, there will be colorful words to attribute to the architect and / or engineer.

SEQUENCE

Given the graphics and the words, the last—and most important— aspect of a successful presentation system is the way the content is used. It is really a matter of organization, and it applies equally whether the job is being described in an hour-long presentation to the client or in a one-minute segment of a slide show for a prospect.

The rule is to use a logical sequence. Translated it simply means: Tell it like it happened.

The client (or any layperson) is just as interested in how the professional arrived at the solution as the professional was while solving the problem. The explanation of this evolutionary sequence is a very essential ingredient of a successful communication system—particularly so when the professional hopes to stretch the client and sell an advanced concept.

The basic rule is: Always begin a design presentation with the beginning:

First, state the problem and enumerate the criteria which govern the design. Sometimes the most effective graphics in a presentation can relate to the program, not the solution. This may be as simple as photographs or drawings (or models) of the existing site, before design began.

Next, show the client *all* the alternate schemes that were explored en route to the chosen solution. (Many think this risks multiple choice by the client, but in practice the opposite is more often true.)

Finally, arrange the order of the presentation so that the recommended solution becomes the logical answer to the problem. This is the surest way to convince the client that there is no better answer.

All this is perhaps easier to say than to do, since it involves an intellectual process. But it has been demonstrated time and again that those professionals who have consistent difficulty persuading clients to accept their recommendations are seldom (despite their claims) dealing with "more difficult" clients. On the contrary, it is usually the professional who creates the problem by overlooking the importance of communicating with the client in a logical, effective manner all along the way.

Professional respect is earned, not won. Learning to communicate ideas on a logical basis is an elemental but sure route to making the point.

At the outset of this chapter it was noted that the professionals who do the most successful designs are also the most effective communicators. It is worth asking whether this is actually true. Could it be that the ability to communicate effectively is a prerequisite to doing the most professional work?

Perhaps it is a little bit of both. If so, developing an effective communication system becomes not just an ingredient of the business development process, but also a big step toward doing better work.

And there is no better sales tool than good work.

from: **the Coxe Letter**

CASE STUDY #28: HELPING THE CLIENT DO YOUR PUBLIC RELATIONS

As its new $11 million addition neared completion, St. Joseph's Hospital, Tucson, Arizona, began comprehensive plans to make the dedication a major community event. One of the early steps of the hospital's public relations task force was to ask the architects, William Wilde and Associates, for help, especially in assembling drawings and other presentation materials.

The architects willingly agreed, and offered the services of their PR consultant to assist the hospital in any way possible. The hospital welcomed the additional assistance and in subsequent weeks called on the architects for:

● Written architectural descriptions of the new building.
● Photographs of sketches and renderings.
● Help in briefing the hospital's photographer on the kind of views to take.
● Written background on the firm and its design process in executing the project.
● An opportunity for the hospital's photographer to take pictures of the architects at work in the firm's offices.

The hospital used these materials in a series of publications issued during dedication week. These included:

● A major dedication brochure with six pages entitled "The Architect/The Plan".
● Six full-page advertisements in the Tucson papers, the first entitled: "The Architect", paid for by contributions from the building contractors.

- A series of publicity releases including a 22″ × 28″ poster/invitation that included in its text 120 words of credit to the architects.

The result is a classic example of how clients welcome, and reward, architects who assist them. "Although we did not seek the recognition" says Richard L. Anderson, AIA, vice president of the Wilde firm, "needless to say we are delighted with the quality of what was portrayed about the architect."

17

THE PROFESSIONAL BROCHURE

"A brochure is a crash program of getting together a bunch of old photographs when someone says they want to talk to you."

Robert M. Coleman, AIA

There seems to be a great deal of misunderstanding about what can be accomplished by having a professional brochure. As a result, a tremendous amount of money is wasted by professionals on brochures which are largely unused and ineffective.

When one does not have a brochure it can seem to be the single most essential element of the marketing effort. Yet once a new brochure is in hand it may appear to be the least valuable tool of the selling system. Understanding this apparent contradiction can help put the brochure in perspective.

In effect, a brochure is like a suit of clothes. Without the clothes of a brochure one seems naked and can feel quite exposed in the marketplace. Once clothed with a new brochure, however, one becomes little different from everyone else who has dressed to be out that day. The cut and style of one's clothes may initially give a prospective client some method of telling two professionals apart, but intelligent clients quickly learn not to judge one from another solely on their sartorial appearance.

In this climate design firms tend to go through cycles with their approaches to brochures, just as clothes go through fashions. Some years ago, everyone wanted a portfolio—a brochure which included just about everything the firm had ever done. Then, as these became too bulky, the deck-of-cards approach evolved using loose leaf pages which can be shuffled to deal different elements in different situations. About the time a majority of firms caught on to this approach, the fashion leaders (or their clients) seemed to tire of the lack of personality provided by the deck of cards, and brochures suddenly became smaller (thinner), briefer, more general, but more distinctive.

As a practical matter, the whole brochure issue is usually given more attention and energy than it deserves. Firms try to "solve" their brochure needs when, in fact, there can never be a lasting solution. Once a firm finds an effective format and clients have seen it a few times and envious competitors have copied it for themselves, that format has served its purpose and it is time to start all over.

It is better to look at the brochure as a short-term statement of what the firm thinks of itself, and to expect to redo it regularly as long as the firm stays in practice. How often? There are no fixed rules, but it seems a good idea to plan for a new edition about every two years. The effort to actually produce a new brochure will take at least a year longer than planned, which means in practice a new brochure will go into circulation every three or four years. (One effective way to speed up the process is to deliberately underprint the estimated need for the current brochure. Nothing speeds up the production of a new brochure faster than running out of the old one.)

As a practical matter, professional design firms need brochures for two very different purposes:

One is the *office brochure,* which tells the stranger in general terms what the professional's practice is all about. This kind of brochure can generally be produced in some quantity.

The other is the *selling brochure,* which must be aimed very specifically at the information requirements of a prospective client. Because no two prospects are exactly the same, the ideal selling brochure should be able to be changed for every use.

To put it another way:

The *office brochure* answers the question: "What do you do?"

The *selling brochure* answers the question: "Can you do my project?"

No professional should undertake production of a brochure without clearly understanding the difference between the two.

The *office brochure,* for example, tends to be historical, biographical, and general. Such brochures can be useful as direct mail pieces for distribution among important contacts and as initial credentials with prospective clients; and they are valuable aids in recruiting and orientation of professional staff.

The *selling brochure,* on the other hand, wants to be adaptable to the information needs of the prospect. Selling brochures tend to be very specific, and their distribution is quite limited. The ideal selling brochure which helps land one particular commission may never be wholly applicable again.

The importance of understanding the difference between the *office* and the *selling* brochures is both practical and economic. Unless it is kept clearly in mind throughout the development of a new brochure, the well-intended practitioner may spend thousands of dollars on 500 copies of a very attractive brochure, only to find three or four years later that two-thirds of them remain unused, they are all out of date, and there is no way of knowing whether they did the firm any good.

It should be emphasized that the effectiveness of a professional brochure has very little to do with how much money is spent. The very best brochure can be a hand-made affair composed of original photographs and typewritten or Ozalid-printed text. Such a presentation is perfectly capable of holding its own against the most expensive product of the printing industry—if the message is there.

BROCHURE FORMATS

The function of the brochure format is to resolve the conflicting requirements of the *office* and *selling* brochure needs in the most effective manner.

There are a number of reasonable format approaches which have become popular among professional firms, and there are undoubtedly fresh answers yet untried. Four of the formats illustrate the current state of the art.

The Credentials Brochure

At this writing a majority of firms are using as their basic brochure a brief, general statement that aims to answer the questions:

Who we are.

What we do (e.g., services).

For whom we do it.

The significant exception in this format is omission of "what the firm has done," in terms of specific projects. By eliminating this and thereby avoiding all the associated questions about what jobs get shown, and in what order, the credentials brochure becomes as flexible and applicable as a calling card—in fact they are sometimes referred to as "calling-card" brochures.

Typical credentials brochures can be in size from 4 to 24 pages and will include:

a. An introductory statement about the scope or history of the firm.

b. A list or presentation of the services the firm performs.

c. Biographies of principals and key staff.

d. A list of past clients, or references.

Although credentials brochures can be wholly text (and sometimes this is very effective), most are illustrated with photographs and/or drawings of the firm's work, but presented without identification as specific projects. Other common illustrations used in credentials brochures are pictures of the principals and key personnel, and of the firm's offices.

The advantage of a credentials brochure is that it is just a suit of clothes. It covers your nakedness and gets you into most any marketplace effectively and economically.

The disadvantage is that it is too general for most selling situations, and therefore must be supplemented when used to pursue a particular project. Many firms have considerable success relying on a printed credentials brochure and custom-made (of loose photographs and typed text) supplements for individual selling situations. Where the custom approach is considered inappropriate, the most common alternatives are the following.

Combination Brochure

This "package approach" is perhaps most frequently used by firms which have made an initial investment in a general office brochure and then learned its limitations as a selling instrument. The combination contains all the firm's experience in one volume, but the project types are grouped (usually between distinctive divider pages).

This type of system is particularly popular with engineering firms whose practice divides into several clearly related categories of work. But it is a compromise whenever it is used to focus on very specific experience, and the prospective client knows it. Also, preparing additional material for each added project becomes a practical burden, and can get quite expensive if typesetting is needed to match special inserts to the original.

Multiple Brochures

A popular alternative is production of several special brochures aimed at the building types which are most frequently sought, such as bridges, hospitals, highways, water treatment, etc. These special brochures are quite effective because they are aimed directly at the prospective client. The system falls short, however, when an unusual building type is being sought and the firm has no alternative but to fall back on its general brochure.

Brochure System

The most common compromise format seems to be a completely looseleaf brochure *system* which permits each copy to be collated and assembled according to the tactics of the particular project being sought. Such a system can even have different covers, with such titles as "Schools," "Buildings," etc. and the order of inside pages can be arranged at will to call attention to the firm's particular experience applicable to the sales situation. This system achieves the same directness as handmade brochures, while the use of preprinted materials makes copies much easier and faster to assemble. Disadvantages are the first cost, virtually all of the firm's past experience and all personnel biographies must be printed and available at the outset if the system is to work efficiently; and limited flexibility in graphic design, because of the requirement for the loose-leaf binding method.

Actually, binding is a design problem in all formats. It is wise to use a binder which permits periodic addition of updated pages without destroying the rest of the brochure. Plastic ring binding does this most easily and cheaply, but has become such a cliché that most design professionals dislike to use it. A few more attractive alternatives are coming onto the market. Although they tend to be much more expensive ($1 to $3 per copy, instead of a few cents) they are worth looking at.

BROCHURE CONTENT

Although format is the first consideration in establishing the approach to a brochure, content is the single most important criteria which will govern its success (and economic return). Ideally, content and format can be so integrated that they become an entity, but this is a rare achievement. If emphasis must be placed on one or the other, content deserves the higher priority.

There are no absolute rules about the content of a professional brochure.

The primary objective is to project the essence of the firm. The more personality the brochure has the more likely it will succeed. But there are some general considerations which should be recognized in framing the content. The most important of these are:

Brochures are scanned, not read. Therefore, important messages should be brief and stand out clearly.

Brochures should state what the firm does. "Architecture," "engineering," or "planning" are inadequate descriptions of the services a professional firm provides. An itemized list will often suggest services the prospective client needs, but has no idea the professional performs.

Pictures are worth a thousand words, but a picture without a caption is merely an illustration, not a sales aid. Captions should describe what the professional did, not just what the structure does.

A bad picture is worse than no picture at all.

In selling brochures (as different from office brochures) the prospective client wants, more than anything else, to know that the firm has experience with the client's kind of problem. Such experience, if possible, should be the first thing seen in the selling brochure.

Design philosophy—whether stated in text or illustrations—is the last thing most prospective clients want to know about. Include it if you must, but briefly. (This does not mean that having a design philosophy is unimportant. It is, of course, essential to doing good work. But the patient cares little about the doctor's medical philosophy—the ill person most wants to know if the doctor will cure the ailment. Similarly, the first thing the building client wants to know is that the professional can solve the problem, not how.)

The quantity of material included in the *selling* brochure is much less important than the quality of what is included.

Brevity is a virtue.

These are the hard-content considerations which will determine whether or not a brochure is an effective instrument. How to achieve them is a wholly different matter that depends in great measure on what the principals who are guiding the brochure process understand about effective marketing communications.

GETTING THE BROCHURE DONE

However the making of a brochure is carried out—whether by in-house effort or with help from outside consultants—it is much easier if the leaders of the firm have a clear idea of how they want to present themselves. If one goes to a clothing store with an idea of what one wants, it is easy to pick over the rack and identify the right outfits. But those who go to a tailor and say "Dress me" without knowing what they want, will spend many agonizing hours trying to pick a comfortable style and will often come out with a compromise that doesn't feel quite right.

There are a number of comprehensive books on brochure preparation (see Bibliography in the Appendix), but all these can do is explain the alternatives and the process of brochure preparation. At some point, the person(s) in the firm with the last word on the subject must decide: "This is what we want to say about ourselves." That direction will define the content, and the contents will make the brochure.

To carry out that direction, however, other disciplines must be considered. In a visual age, there is a clear correlation between the graphic quality of a brochure and its impact on a marketing audience. At the same time there is also a distinct line between graphic designs which portray professionalism, and graphics which look commercial or contrived. Thus, the approach to design of a brochure is a very important consideration.

Achieving the best graphic solution first involves consideration of who will do the design. Generally engineers look to outside consultants for help, while architects try to do it themselves. There is good and bad in both approaches.

The key to successful graphic design of a brochure is to work with a designer who can project your essence, rather than the designer's idea of what you ought to look like. An in-house designer is clearly ahead on this score, but often they have only limited understanding of printing technology and either miss opportunities or overcomplicate the process. Outside designers, on the other hand, must be chosen with considerable care or your brochure will look more like their work than like your firm.

It is good practice for a professional firm to select and educate a graphic design consultant with the same care that one would want to select (or be selected as) an architect or engineer. This means preparing a written program, calling for paperwork responses from a long list, interviewing a select few, and then investing enough in the chosen designer so the consultant becomes part of your marketing team. Far too many professional firms consider this selection process too much effort, and their brochures show it.

The individuality that all firms should strive for in their brochures (and other marketing tools) will only be achieved if the firm itself is prepared to give the substantial input required. That usually is 60% to 80% of the total effort. The total that can be contributed by outside help, in graphics, writing, and possibly some marketing consulting, is only 20% to 40% of the whole, but may make all the difference between the result being average or outstanding. At the very least, when outside help is involved, the brochure generally gets done more quickly.

Getting a brochure done inevitably seems to take longer, cost more, and involve far more complications than firms expect when they go into the process. The reason is usually simple: Firms simply underestimate the number of policy decisions required to establish how they want to present themselves. If these can be decided upfront, the brochure process becomes more manageable.

BROCHURE COSTS

The decision on how much to invest in a new brochure should be a straight value judgment related to how much time and effort the firm intends to devote to business development and what returns are expected. There are four major elements which make up the cost of a professional brochure, and they break down roughly as follows:

35% Concept development, data gathering, writing, and supervision.

30% Illustrations (photography and/or drawings).

15% Graphic design.

20% Production (typesetting, printing, and binding).

The budget for *production* of the brochure (as opposed to preparation cost) should be based on the useful life of the material and the quantity to be put to use. A moderate-size office which lands six or eight new clients a year may only use a dozen *selling brochures* every 12 months—and half of these may be recoverable. If handmade brochures are used, the material cost for original photoprints, etc. plus staff time may be as much as $200 per copy, but this adds up to a total brochure budget of only $600 per year. Suppose on the other hand the firm prints 500 copies of a brochure at a production cost of $7500—or $15 per copy. If no more than a dozen a year are used for *selling* purposes each year, and if the edition is obsolete in four years, the effective cost of having the printed $15 brochures may be more than the cost of using handmade ones. The only real advantage is convenience.

The payoff with printed brochures, therefore, increases greatly when the quantity used goes up. This speaks in favor of investing first in a printed *office,* or credentials brochure, and using handmade selling materials until the volume or efficiency makes printed materials worthwhile.

SUMMARY

The foregoing reviews the considerations which govern development of the professional brochure. It must be stressed again, however, that having a brochure is not the same thing as having a viable business development program. For this reason, the young firm, using handmade materials and perhaps a 4-page credential costing less than $1000, can have just as effective a brochure as the firm which invests $20,000 to $50,000 in a whole brochure system.

The brochure is just a tool in the selling process. The sharper the tool, the better job it will do. But without someone to wield the tool, it will do no work at all.

from: **the Coxe Letter**

CASE STUDY #67: SHORT BROCHURE GETS RESULTS

Alberta, Canada, design firm Cohos, Evamy & Partners had been using a conventional 12-page illustrated brochure format for 10 years and was ready for a change. "It looked good but nobody would open it," says partner John Zuk.

A little over a year ago, Zuk got the client's eye view of office brochures when he was choosing a special consultant. "I didn't want to go through their brochures. They were trying to glorify themselves, and all I wanted was the meat."

As a result, Zuk designed a simple, 4-page, 8½" × 11" brochure. On the front is a statement of philosophy and capability. Page 2 lists key people and membership affiliations. Specific services are listed on page 3, under the categories Architecture, Structural Engineering, Mechanical Engineering, Electrical Engineering, Municipal Engineering, Urban and Regional Planning, and Interior Design. The back page has a partial list of clients. It was printed in one color on cover stock at a cost of $550 for 1500.

The brochure was initially mailed to about 350 past and prospective clients. When the firm is seeking a specific job, a personal cover letter addressing the particular project is added, and suitable photographs may be enclosed.

"The response has been very positive," reports Zuk. "People are saying, 'It's all I wanted to know.'" A long-term prospective client said, "I thought you only did architecture; I didn't know you were engineers, too" and promptly gave the firm a job. Even existing clients have shown greater awareness of the firm's services—a municipal client saw the new brochure and said, "I didn't realize you did interior design. I have another project for you."

CASE STUDY #78: AWARD-WINNING BROCHURE SPURS SELLING CONFIDENCE

In order to plan for the eventual retirement of its founding partner, the Houston architectural firm of S. I. Morris Associates committed itself to a marketing program in which all partners would take part in the business development contact effort. The firm's marketing coordinator, Janet Goodman, developed a list of organizations and people for each partner to contact. Many of the partners had never called on prospects before. "They were receptive to the idea, but they didn't know how to break the ice, and consequently many calls were not being made."

"We thought a new brochure that reflected the caliber of the firm's work would help," she adds. A committee was formed, composed of Gene Aubrey, FAIA, Chief of Design; John Wiegman, AIA, Director of Business Development; and Goodman. The group hired Dan Glidden of Flat Lizard (sic) Graphics, for outside assistance. The fact that he'd never done an architectural brochure was considered an asset—the committee wanted "something that would distinguish us from other firms."

"We let him direct us," says Goodman, "and asked only that he not allow us to make any mistakes." They met with him weekly for three months. Glidden recommended and directed photographer Ron Scott, who took 90 rolls of film on 8 selected projects. Glidden criticized the first text, drafted by the committee, for being "too architectural" and got a professional writer to rewrite the copy. After several edits by everyone involved, Glidden gave it the nod, saying that finally it wasn't "architects talking to architects".

The finished product is a 24-page, 4-color brochure. The first page, "Solutions", describes the firm, its philosophy, and its services. Following it are seven double-page spreads, each summarizing a single project in 100 to 150 words with two or three dramatic photographs. The final page is a partial list of clients. Five thousand copies were printed. Total cost, including design, photography, and printing, came to about $25,000.

Nine hundred were mailed to a carefully-developed list of community decision-makers and past clients with a short cover letter signed by the appropriate partner. "The response has been overwhelming," says Goodman. The firm has received 120 complimentary letters. The head of a large international firm responded, "Regrettably we have no projects at this time, but when we do we'll call you." A local developer walked into the office without an appointment and said: "You did all the best work in town. Can you really give a 'quick conceptual response' like your brochure says?" When the firm assured him it could, he commissioned them on the spot to master plan a $100 million office complex.

As soon as the brochure came off the press, the 16 partners began making personal visits to the people on their lists, covering about 60 people a month. Says Goodman, "They feel the brochure strongly illustrates the firm's ability to do work for prestigious clients. They know that a brochure doesn't bring in jobs—that's their assignment. But this brochure enhances the credibility of what they say. Some of them who had never made a sales call before are now making major presentations.

18

PUBLICITY

"Don't feel you have to take all those media people out to lunch. They don't want to eat with you any more than you want to eat with them—they're just looking for good stories."
Pat Harris, V.P. Public Relations, Gerald Hines
Interests

Publicity is by far the most misunderstood communication technique in the professional world.

For a variety of reasons, a great number of professionals seem to feel that publicity is the key to success of their competitors, but that the media are in a conspiracy to deny all other professionals their rightful recognition.

Let it be said forcefully for all to understand: *As far as professionals are concerned, publicity is always the result of success, never the cause of it.*

If this can be accepted, then it will be no surprise that what one reads in the papers is usually about the most successful professionals. They attract publicity because they are successful, not vice versa. Frank Lloyd Wright never had a press agent, but after he became famous no one ever forgot to print his name in a newspaper article about his work.

There are several other axioms about publicity which every professional should understand.

No professional ever became successful because of publicity.

Many very successful professionals have never in their careers sought or received publicity.

Except in rare instances, publicity does not produce commissions.

Publicity can be useful in a business development program, but for the great majority of firms it is the least important ingredient of the selling process.

Some will argue that there are examples where the above do not apply. There are exceptions, of course: For example, publicity can be valuable to an architect specializing in custom houses. But remember, the first house had to be built to get the publicity, not vice versa. If publicity is to be useful at all, the professional must first understand and accept the realities of it.

PUBLICITY OUTLETS

Within the larger context of the generalizations made above, there are very great differences in the publicity media which can be useful to the professional.

Local Press

It is not really an oversimplification to state that the greatest value a professional can gain from local newspapers, radio, and television is what can be learned from them about prospective clients, not vice versa. The local press reports *news*. The news which a professional generates and which might be of interest to local press is such things as:

Selection for major commissions, particularly public work.

Completion of design—the rendering with an announcement of plans for the project, plus the graphic rendering.

Award of construction contracts.

Groundbreaking.

Unusual construction aspects in progress—particularly when visible to sidewalk superintendents.

Opening of the completed structure.

Receipt of recognition or awards.

It is worth noting that, if a story about each of these events were published, only the first and the last would require that the name of the professional be mentioned. In this type of story the design firm is the news. In all the other stories the basic news relates to the owner, the contractor, the occupant, etc.

When a rendering is published in a newspaper the news is what the building is going to look like. Who designed it is generally quite secondary to community considerations, as with a public building; or business considerations, as with a new plant, office, or store.

Therefore, if the architect or engineer wants publicity about these aspects of their projects, they must take the initiative to get it. Local publicity will not come to a firm. No matter how much architects or engineers may think they are the news, it is not so.

More of this later; for now it is important only to remember where the local press fit into the publicity media.

Trade Publications

There are more than 4000 listed trade publications published in the United States. They exist because their areas of concern are too limited for local press or mass media yet they report information for which there is a definable audience. Examples of such publications:

College and University Business
School Management
Your Church
Factory
National Real Estate Investor
Apartment Construction News
House and Home
Nation's Cities
Hospitals

These publications have a natural interest in the work a professional may do in their particular field. Such articles will, of course, relate mostly to the project or the client, for it is the professional's clients who are the readers.

Publicity in these journals can be of obvious value to a professional, but if one wants coverage in such publications, one will generally have to take the initiative to get it.

Professional Magazines

The architectural, engineering, and related journals comprise that segment of the trade press that wants to communicate with professionals. Here the professional's name always gets in print because that is the news. These publications will seek an architect or engineer whose work is at all unusual or distinctive. But the readers of these publications are other professionals, not prospective clients. The professional magazine audience is important to larger firms which want to recruit talent and it yields a certain standing among one's peers which may lead to referrals. But *per se,* articles in professional journals will not normally reach prospective clients unless they are reprinted and circulated by the professional himself.

PUBLICITY RELEASE AND PLACEMENT

Within the context of the media described above, it must be also recognized that the ratio of available news to the total of print space or broadcast time is increasingly out of balance. As our society and communication capability has mushroomed, it is no longer possible for any publisher to pretend to cover "All the News that's Fit to Print." Editorial space in publications is related to the volume of advertisements which can be sold, and reporters are hired in relation to the number of columns of copy they can produce in a given time. Publishers don't hire reporters to gather news which they don't have space to print. Thus it is quite possible to understand why, if there are six groundbreakings in San Francisco on Monday, the one which gets into print is that which is handled by the ablest publicist. The reporters haven't time to cover any of them, and the editors don't have space to print all six even if they get six well-done releases.

Similarly, *School Management* magazine and *Architectural Record* do not expect to print every example of a solar heated elementary school in an article on the subject. If there are 10 examples in the country, the three which get into print are likely to be the three which arrive on the editor's desk complete with the best photographs and descriptive texts.

The moral in today's publicity world is: The Fourth Estate helps those who help themselves.

This does not mean one needs to pay tribute or hire magicians to receive publicity. Editors are just as ethical about their professional standards as architects and engineers are about theirs. Thus, most editors welcome releases of legitimate news, and evaluate such material for publication without bias.

There are two basic approaches to self-help publicity:

The news release is a prepared story which is reproduced in quantity and distributed to all outlets which might be interested in the story. Releases of this sort are advantageous when one hopes to have the item appear in a number of outlets. The disadvantage is that editors who receive such releases can assume that all competing publications have received the same thing, and therefore may discard the story thinking someone else will print it.

Direct publicity placement is a technique whereby the news source selects one specific publication, approaches the editor to determine interest, and then prepares the material to suit that publication's needs. The advantage of direct placement is that the probability of publication can be determined before investing the effort in preparation, writing, and illustrations. The disadvantage is that editors who accept stories by direct placement generally ask for exclusives, making it difficult to receive exposure for the same story in more than one publication.

In practice the choice between the two publicity approaches is determined by the timeliness and complexity of the material. From a sales standpoint, however, the publicity that is usually most valuable is articles which can be reprinted and distributed in the selling process or by direct-mail (see Chapter 19). For such articles to be worthwhile sales aids, it is generally better to have them specifically placed rather than broadly released.

A special complication is the practice among the major architectural magazines of publishing feature articles *only* on an exclusive basis. Thus, if *Architectural Record* publishes an office building, *Progressive Architecture* will refuse to carry an article on the same building. In view of this, publicists for A/E firms will often shop around among the design magazines to be sure of the best display before accepting a "commitment" for publication from one journal. The practice has the merit of assuring publication among the journals of the largest number of different projects. Its drawback is that once the professional "commits" a building to one magazine, the selected publication may sit on the article for months, and in the meantime the professional may lose more favorable opportunities for publication. Other trade publication fields where there are competing magazines operate on a first come, first served basis. Exclusives can and are given, but it should be on a *quid pro quo* basis (e.g., the publicist will agree to an exclusive release in exchange for a "cover story" treatment).

HOW TO DO IT

From the above it should be obvious that professionals who want publicity must be prepared to do something to get it.

There are three ways to go about it:

1. Do it yourself. This requires telephoning, visits, and correspondence with editors; learning their requirements; preparing the necessary material; and developing a distribution list. It is perfectly proper form to do it yourself, but budget a lot of time, and ask whether the time would not be better spent in direct contact sales efforts.

2. Engage a publicity specialist. He or she will do exactly what you can do, in a lot less time, and with fewer wasted efforts.

3. Let your clients do it for you.

For general publicity, the last alternative is by far the most effective and when combined with the input of a publicity specialist can be extremely efficient. Virtually every major client today has its own publicity needs and systems. Usually the client has more reason to want the project publicized than the architect or engineer—to raise financing, find tenants, earn community relations, etc. In fact, most of the project publicity which gets in print these days is put there by the clients. Since they are already doing it, the architect or engineer is actually doing them a favor by seeking publicity at the same time.

It works this way. When the client wants to publicize the proposed design, groundbreaking, etc., the client's public relations staff will need facts, illustrations, etc. If this information is developed from the client's facilities department or the general contractor, there is a good chance the publicist will never learn who the designer was, let alone why a good quote about the design should be in the release. Such information is not essential to the client's story needs, so the release goes out without it.

If, on the other hand, the architect or engineer anticipates this, the client can be asked in advance whether there will be publicity about the building, and the designer can offer to help. Assuming also that the professional has a good presentation system (Chapter 16) in operation, there will be ready-made materials to give to the client's publicity representative. The professional's own publicity representative can even volunteer to draft the release; it is then distributed on the client's letterhead.

It is just as simple as that. By becoming involved in the client's publicity, any architect, planner, engineer, or other professional can get work in print almost every time.

This basic principle has been expanded by many of the larger A/E firms to the point where they publish financing, leasing, or dedication brochures and place publicity *for their clients,* and get paid for it in the bargain. It is called marketing support service, and the firms which offer it sometimes get hired because of it.

A case study illustrates how effective this approach can be. An architectural firm was commissioned to design a medical school. The facility, new from the ground up, was the first of its type in the nation and it was located in a community where no similar institution existed. The whole project was newsworthy, and the press was very interested in it. The client was publicity oriented and opened a PR office on the site well before the buildings began to rise.

However, throughout the initial planning and during construction of the first stage, the architectural firm never found its name in the client's local press releases or in national trade articles authored by the institution's staff. In several instances the architectural firm had been asked to provide special reproducible floor plans and model photographs to illustrate these articles, but still it received no credit. Eventually, the firm concluded the client was in some manner dissatisfied with the firm's performance and was deliberately denying the firm recognition.

As the time neared for dedication of the first stage of buildings, the architect tried a new approach. The architect was being asked to supply plans to give to the visitors. Aware that the facility was being toured regularly by important groups from the medical world, and realizing many of these visitors represented potential clients, the architect suggested to the client that a brochure be published explaining the design and including sketch plans.

The client was delighted and accepted the architect's offer to help. The architect's public relations representative was put in touch with the client's PR director who welcomed professional assistance. An arrangement was worked out whereby the architect designed, wrote, and illustrated a 24 page brochure (out-of-pocket cost $2000) and the client paid to print it (cost $3000). In executing the brochure, the professional firm had the opportunity to see that the architectural concepts were appropriately highlighted. The firm's name was prominently included in the credits.

While the brochure was being produced, the client's PR director asked for the architect's assistance in planning for a ceremony to dedicate the initial structure. The architect took the leadership, and among other things, offered to provide an engraved silver trowel for use in setting a datestone at the ceremony. In return, the principal of the architectural firm was invited to participate in the formal ceremony together with the governor of the state, the president of the medical school, and the surgeon general of the United States. Of course the architect's name was also prominently included in the program for the day and in the releases in the press kit.

The morning after the dedication, the state capitol newspaper carried a photo of the "distinguished guests" at the ceremony, including the architect, and his name appeared prominently in the text of the accompanying front page story.

All of this came about with a client whom the architect had seriously thought was conspiring to deny him publicity!

By taking the initiative and *offering to help the client meet its own publicity needs,* the design firm had:

1. Gained control of the content and design of the facilities brochure, thereby assuring its appropriateness from the professional standpoint.

2. Gained recognition with an audience in the medical education fraternity, through the client's distribution of the brochure.

3. Obtained (by printing 500 extra brochures) an excellent sales piece to use in the firm's direct mail program and in answering subsequent inquiries from prospective clients.

4. Received public recognition at the opening ceremonies before a VIP audience.

5. Got the firm's name in the newspapers.

6. Earned, by being helpful, the lasting respect and friendship of the client's PR department, thereby assuring continuing cooperative involvement throughout the life of the project.

This is the *client-oriented* approach to publicity. If a professional firm wants to spend time and/or money on publicity, this is the place to start.

from: **the Coxe Letter**

CASE STUDY #15: GETTING IN A LOCAL NEWSPAPER

Eight months ago, Frank Lamb, AIA, telephoned the Orlando Sentinel/Star and asked the real estate editor for an appointment to talk about the work of his 5-man architectural office. His objective was to seek more attention for his six-year-old practice, so when he met with one of the newspaper writers he asked:

"What are you looking for?" "How can I give you information in a form that is most useful?"

The reporter explained that the paper was interested in residential projects and welcomed suggestions of projects that are "a little out of the ordinary" and might suggest ideas to a typical homeowner. He told Lamb that a phone call or visit is all that was needed to have an idea considered.

The results of that brief introduction have been dramatic. Five half-page articles on Lamb's work have already appeared, and several are in the works. The subjects initially were individual residences (such as one over a stream) but as Lamb and the reporter visited these projects they discussed broader subjects. One conversation about avoiding solar heat gain led to a meeting at which Lamb sketched out the format for a house with solar screening and free air ventilation concepts ("Principles every architect in the country knows," says Lamb). The paper printed the sketches and a long article. The day after that article appeared Lamb received dozens of calls from people who wanted plans. (He turned over the calls to a stock plan drafting service in which he is part owner, and 15 sets have since been sold at $150 each.)

Lamb cannot trace any other direct work to the publicity, but he is finding what he considers much more important benefits.

"The publicity is making a tremendous difference in how hard we have to work to sell a design idea" to existing clients (developers, a medical office group, several churches), reports Lamb. "Apparently, if your name is in the paper you are supposed to know what you are talking about. It's just amazing."

Most valuable, Lamb feels, is the relationship of mutual interest he has established with the newspaper. "They welcome a source that can help them," Lamb says. "Now we no longer have to call them—they call us."

CASE STUDY #77: PUBLICITY PAYOFF

PKR Associates of Milwaukee was commissioned by St. Francis Hospital to remodel a substantial portion of the maternity ward. The design incorporated some unusual features, including a "new parents" dining area. "It was a relatively new concept," says firm vice-president Eugene C. Schmitt. "It provides table service in a cabaret atmosphere where new fathers can join their wives for a leisurely dinner."

Jack Funck, project manager and head of PKR's interior design department, thought that press coverage would benefit the hospital and could also show the firm's capability in renovations. And he believed the project had the dimension of human interest that would appeal to a newspaper. So he called up a *Milwaukee Journal* reporter, one he knew to be knowledgeable about architecture. ("I research my reporters," he notes. "I had seen this one at AIA meetings and that impressed me.") "The reporter was interested, and I took her to see the project. She interviewed me while we were walking around," says Funck. "I explained what the problem was and how it was solved."

The story ran on the front page of the Sunday "Home and Real Estate" section, with a color photograph.

The weekend the article appeared, Richard R. Barr, the Administrator of Albuquerque's Presbyterian Hospital Center, happened to be visiting in Milwaukee. He clipped the article, flew back to New Mexico on Monday, and phoned PKR that afternoon. We're doing a maternity ward interior," Barr told Funck, "and we'd like to talk to you." Two days later Funck met with him and the Center's Board. Funck went home with the commission.

Since then, PKR has been hired by the same client to remodel several other hospital floors, then again for more remodeling. Subsequently a group of doctors asked PKR to remodel their offices in a nearby building.

Says Funck, "I think newspaper articles are a good way of letting people know what you do. About 40% of those we've had resulted in some kind of call to us."

19

ADVERTISING AND DIRECT MAIL

"I don't know who you are.
I don't know your company.
I don't know your company's product.
I don't know what your company stands for.
I don't know your company's customers.
I don't know your company's record.
I don't know your company's reputation.
Now—what was it you wanted to sell me?"
Famous ad for McGraw-Hill Magazines

It took the design professions years to accept John McGraw's wisdom about marketing in a complex society. But now, with an assist from the U.S. Justice Department (see Chapter 2) the rules allow architects and engineers to apply common sense in their use of modern communication media and techniques in support of their marketing efforts.

It is interesting to note that since the professional societies revised their rules about advertising, there have been no immediate, wholesale changes in the way design firms are merchandising themselves. To date, the experience of those who have experimented with broader advertising techniques falls into two categories:

A few firms that practice in defined markets or offer specialized services feel they are getting results from selected space advertising in trade and regional publications.

Many more firms of all sorts are using direct mail to reach past clients, contacts, and identifiable marketing audiences.

This is no surprise. As discussed in Chapter 5 the key to all effective communication is knowing your audience. Most architects and engineers can be quite specific about their audience, and any advertising medium which reaches them deserves consideration. Direct mail is the obvious first choice in most cases. Space advertising can also deserve consideration in those cases where the audience is sufficiently clear.

The state of the advertising art in marketing design services will undoubtedly evolve for some time, but there is no longer any doubt about whether it has value to a professional practice. The only real questions now are where to use it and how much to do.

Some firms by the nature of their practice need little or no merchandising and would be unlikely to get value from an advertising investment. These are primarily practices that rely on a few existing clients for their work, and are doing no other active marketing.

The majority of firms which rely heavily on reactive selling to opportunities which come to their attention (as different from firms doing a lot of early birddogging and courting well before targets are in sight) may find direct mail of primary value as a means of efficiently keeping in touch with past clients and contacts.

The firms that will be able to make the most extensive use of both space advertising and direct mail will be those which are most actively prospecting among new clients and in new markets. The ability of targeted promotion to open doors for these firms will make the effort and investment well worthwhile.

SPACE ADVERTISING

Effective use of space advertising requires a number of disciplines—e.g., copywriting, graphic design, media placement—which are not normally found in-house in design or other business firms. This is what advertising agencies are constituted to do, and as a practical matter it is not wise to attempt a space advertising campaign of any scope or duration without the help of such an agency. This advice, however, creates a problem for many design firms because the amount of advertising even the larger architectural or engineering firms might budget is seldom an attractive account to the best advertising agencies. This fact alone may limit the use of advertising by design firms.

Those design firms which have experience using advertising agencies give advice that sounds remarkably similar to what clients for architectural or engineering services might say about the selection of design firms. The major points:

1. Select an advertising agency only after serious interviews with a number of interested firms.

2. Be absolutely clear who will be the account executive and who will be the copywriter, and be sure you are comfortable that they can learn to understand your services and communicate on your wavelength. (For example, successful experience advertising consumer products, business equipment, or even building materials may be no preparation for successful advertising of professional services.)

3. Be prepared to pay the agency for its time, not just a commission on the advertising it places. Most design firms don't require anything like the amount of space that would be required to compensate the agency through commissions. In many cases the agency fees may well equal or exceed the cost of the space purchased.

4. Once an agency is selected, expect to spend substantial time educating them in the goals and marketing objectives of your firm, and thereby learning to work together.

The last point in particular may restrict many design firms from using space advertising. Getting a good advertising program underway takes so much internal effort that it should not be looked at as a short term experiment. The total time to choose an agency, design a program, and execute it long enough to judge the results is seldom less than 18 months to two years.

All these caveats aside, there are a growing number of cases where design firms have created effective advertising which they have found worth the effort. Some of these cases have been published (see Bibliography), and others can be learned about by talking to the professional societies or directly to other firms.

Some of the more creative examples that have been tried include:

An architectural firm that advertises on the financial page of a major city newspaper whenever it receives a commission from a major corporation, thereby gaining recognition for the client company it keeps.

An engineering firm that sponsors public affairs programs on public television, in order to put its name in front of the civic leader audience reached by the program.

A small, design-oriented architectural firm which advertises philosophical statements about architecture in the concert programs of the local symphony orchestra, in order to associate itself with the cultural audience in its community.

These are only a few examples. Many other firms are advertising in trade journals that reach the audience for their particular specialty (e.g., *Water Pollution Control, Hospitals,* etc.). All such advertising has the potential to be helpful to a marketing program whenever the cost and effort invested is in balance with the return expected.

When that equation is hard to balance with space advertising, the alternative that more often yields the most effective return is direct mail.

DIRECT MAIL

Direct mail is probably the most effective avenue of active merchandising open to the professional. All it requires is:

a. A mailing list, composed of names of individuals with whom the professional has had and/or wants to make contact.

b. Mailing pieces—such as newsletters; new editions of the firm's brochure; project reports; reprints of trade magazine articles about the firm's work; or copies of facilities brochures published by the firm's clients (Chapter 18).

c. Transmittal letters.

Don't associate the type of direct mail program suggested here with "junk" mail. It is quite the opposite. Professional direct mail is generally sent to individuals known personally to the sender or known to be interested in the services the firm offers. Such people usually welcome informational mail from professional firms.

The value of a direct mail program to the professional is the opportunity it affords to inform contacts and others about activities which they might not otherwise know. For example, consider the director of a hospital who is on an architect's mailing list as a former client of the firm. The architect mails the hospital director a trade magazine article about a college library designed by the firm. The hospital director is also a member of the board of his alma mater, which is considering a new library. Ergo, the architect's name goes on an interview list (or the director may phone the architect offering an important lead).

This is the primary purpose of any professional direct mail program: to encourage repeat and referral business. Since these two sources provide 60% to 90% of every professional's business, the value of keeping in touch with past clients, contacts, and friends becomes obvious. Of particular importance is the need to inform these contacts of activities which are outside the normal expectation of the acquaintanceship. In the example above, the hospital administrator might normally consider the professional a hospital specialist. He or she may never know that the professional also is skilled in design of schools, colleges, office buildings, research laboratories, etc., unless the professional takes the trouble to communicate the information. The other value of direct mail is to reach potential contacts which the firm intends to bird-dog in the future. There is no substitute for the response received when one knocks on a strange door and is told: "I've been reading about your firm. I'm glad you stopped by."

Once the value of direct mail is accepted, there is no trick to having a mailing program. A few basic considerations, however, are worth discussing.

The Mailing List

The most valuable mailing list is that of people you already know, and the time to prepare such a list is before it is needed. A few rare individuals are disciplined enough to carry 3 x 5 file cards with them and note down at the end of each day the name and address of any new contacts they have made. Others can accept a system which requires them to fill out weekly a mailing list form along with their time card. Most professionals, however, will have to rely on their secretaries to develop a list by recording those with whom the professional has correspondence or appointments. The secret is to be sure the secretary knows this is a responsibility and is given a card filing system on which to maintain the mailing list. The marketers in the firm may also use the form shown in Figure 1, Chapter 9 to add contacts to a mailing list. The point is, generating the best mailing list is a continuous process. Once the firm makes a discipline of building a list, it is amazing how fast it will grow.

For practical purposes most mailing lists should be cross-indexed by individual and organization:

a. The individual entry should carry the name, title, organization, address, salutation (if first name is used), and the name of the person within the firm who will sign letters addressed to that individual.

b. The organization entry (preferably on another color index card if manually filed) lists the names of all individuals in the organization who are carried on the list. This makes it easy to check whether all the right people in one organization are actually on file.

Size of the list is not a major factor. A good list of 50 names can be more valuable than a bad list of 500. In practice, most firms will find mailings range between 50 and 300 copies if their list is properly prepared. Modern word processors with computer memories can easily accommodate such mailing lists and make updating especially easy.

Mailing Pieces

The choice of media for mail distribution is becoming more important as more and more firms use direct mail in their marketing programs.

Many firms today are publishing quarterly or semiannual newsletters as their primary direct mail vehicle. Newsletters have the advantage of establishing a repetitive format for presenting information you want to have reach your audience, and they are relatively efficient to produce and don't require a lot of original decisions once a format is adopted. Their chief disadvantage is their impersonality—they are inevitably received as broad scale mailing pieces and may not capture the personal attention of individualized direct mail programs.

Individualized mailing programs use custom-typed letters or personalized cards to transmit a variety of elements, including new editions of the firm's brochure; copies of facilities brochures produced by or for clients; and reprints of trade magazine articles or news clippings. Value judgments about where to apply publicity and client relations efforts become a great deal more realistic when a direct mail program is a coordinated part of the firm's business development program. For example, an engineering firm which feels it is overloaded with highway work and wants to expand into industrial structures, can easily judge what sort of mailing pieces it should generate.

In actual fact, the greatest value of a trade article in *Factory* magazine may not be the magazine's circulation at all. But reprints of the article, sent to a few dozen key people on the professional's mailing list, can put it directly in the hands of those who may one day apply the information to the professional's benefit.

The one self-produced mailing piece which is always appropriate is a Christmas card. A number of professional firms have created original cards for years, with the result that the cards are anticipated and talked about. This is somewhat removed from selling but it is valuable public relations.

The Transmittal

The surest way to prevent a direct mail program from being received as junk mail is to send each piece with a personal letter of transmittal. Even if it is the same letter to several hundred people, it should be individually typed (there are word processing services which do this quite economically if you do not have the capacity to do this in-house) and personally signed. Transmittals of this sort accomplish two important objectives:

First, the professional is encouraged to personally screen the list and select those to whom the particular mailing should be sent. Quite often, a given mailing piece will not be appropriate for everyone on the list. Individual transmittal assures an interested reception, and avoids the look of scattershot.

Secondly, the personal transmittal conveys a wholly different image to the recipient. It is not unusual for the recipient to respond with personal acknowledgement. (In some cases of highly appropriate direct mail by professionals the acknowledgement ratio can be up to 50%.) There is no better index of the validity of professional direct mail than complimentary reactions from one's friends and contacts.

Frequency

There are no fixed rules governing how often a professional should circularize the firm's mailing list, but most firms find two to four mailings a year to any one individual are sufficient. Mailings of frequency greater than four times a year should be undertaken with great care to avoid conveying the wrong kind of commercial image.

It should be obvious by now that there is little point in having a direct mail program unless one first has a presentation system which creates suitable materials about one's work (Chapter 16); then uses this to generate publicity articles and client facility brochures (Chapter 18). In this context the direct mail program is simply a downstream adjunct of the total communication system.

This is the real role of the tools of the selling trade. Individually, these communication tools are of only limited value and cannot be relied upon to yield significant returns. But when combined into a total marketing program, they can go a long way to help sell professional service.

from: **the Coxe Letter**

CASE STUDY #24: USING THE XEROX AS A MARKETING TOOL

Long Beach architect Kenneth Wing, Jr., AIA, has a highly personalized mail out program that reaches 75 to 100 clients and potential clients each year at a cost he estimates at about $15 for materials, plus postage.

As Wing reads the technical magazines that come into his office, or the daily newspaper, or airline magazines while traveling, he keeps an eye out for items that might interest someone he knows. "When I run across something" says Wing, "I Xerox the clipping and send it along with a note."

Recipients are "most appreciative," he reports. "In about 40% of the cases I'll receive a phone call or letter saying 'thanks'."

"I couldn't say I've gotten a job from this," says Wing, although his eight-man firm is currently working for several people he once mailed to. "But you must keep your name in front of people. It's my way to say: I'm thinking of you."

CASE STUDY #36: DIRECT MAIL ANNOUNCES A NEW SERVICE

When architects Hawley & Peterson of Palo Alto, California, decided, after a year of deliberation, to add interior designer Catherine Newman to their staff, there was no question in partner William Hawley's mind how the public announcement should be prepared. "I've come to the conclusion that we as architects should stop trying to write our own brochures," he said. "Architects tend to treat brochures as another extension of their architectural service, but they can't do this kind of work as well as a specialist."

In keeping with this conviction, Hawley called in award-winning graphic designer Sam Smidt, who had done the firm's stationery and logotype, and asked him to help produce a brochure announcing the firm's interior design capability.

Smidt initially got together with the partners and Ms. Newman and brainstormed the piece. "There's a lot of our input in the results," says Hawley. Then, Smidt wrote and designed an eight-page, 8″ X 8″ announcement which begins:

"Now that Catherine Newman has joined our staff, the architectural and interior planning for your office remodeling can be fully coordinated under one roof . . .

Ours"

The brochure then contains a photograph and biograhy of Ms. Newman; an offer to show examples of the firm's work and arrange visits to completed projects; and concludes with the firm's new name: Hawley & Peterson, Architects and Interior Planners. The firm ordered 500 copies at a total cost of just over $1000, including the designer's fees and printing.

To distribute the brochure, Hawley reports, "We started down through our Christmas mailing list and came up with about 250 names of important contacts, past clients, and prospective clients." The reaction was immediate. "Each of us got a dozen or more calls from past clients who complimented us on the brochure and the new service."

There were tangible results as well. One prospective client, who the firm had contacted but not previously worked for, hired Hawley & Peterson's Interior Planning Department to do some tenant work in an industrial park. In addition, a former client commissioned them to make a number of changes in a year-old county office building which the firm had designed.

Hawley considers the announcement responsible for both projects as well as for a good deal of subsequent interest in the firm's new service. "We needed to make a strong statement to these people that we were really in the interior design field," he says. "The brochure did it."

CASE STUDY #104: DIRECT MAIL PUTS NEW FIRM ON THE MAP

"McMurray Abernathy Poetzsch/Architects & Planners is 46," says the announcement used by architects Charles L. McMurray, Robert D. Abernathy and Paul B. Poetzsch to announce the start of their new practice in Charlotte, NC. The text refers to their 46 years of combined experience to back up the claim.

Fold-out flaps include informal portraits of the partners, together with their comments on design philosophy and technique. Edited by partner Poetzsch, the thrust of the text is not to belabor the importance of good design. Rather the suggestion is that the firm's integrity is founded on the partners' commonly-held beliefs about architectural practice.

McMurray • Abernethy • Poetzsch/Architects + Planners is 46

Charles, Don and Paul have a combined 46 years experience in the architectural profession, and are pleased to announce the formation of their new firm:

McMurray • Abernethy • Poetzsch
Architects + Planners
830 Providence Road, Charlotte NC
704/372 1860

The announcement was mailed to a list of 1,800 past clients, prospects, and professional associates culled from the partners' experience. Response was "better than we expected," Poetzsch reports. There were about 100 phone calls expressing congratulations and/or support. A similar number of letters, most of which commented favorably on the announcement itself, were also received.

A priest in Charlotte called the day he received the announcement to ask if the firm would be interested in competing for a $300,000 church project; an official of a local university called to invite a proposal for a $4.5 million housing job (he expressed hope that the firm had enough work on the boards, since funding for his project wouldn't be available until July); and the client for a $7 million housing project in Greensboro intimated that the formation of the partnership, as portrayed in the brochure, had been a positive factor in the award of the work (the job was being pursued by one of the partners when the partnership was formed).

SPECIAL REPORT: NEWSLETTERS

In a recent Coxe Letter Market Scan, we asked readers to send us samples of the newsletters they publish and some information about them. Out of 113 architectural firms responding, 25% said that they do not publish a newsletter of any kind and 32% reported they are "thinking about it." A total of 49 firms (43%) publish at least one newsletter—18 have an in-house newsletter, 37 circulate a client newsletter, and 6 reported having both. The Coxe Letter reviewed 20 "internal" newsletters and 35 "external" newsletters. Here are the results:

Newsletters for clients and friends ("external"). Firms responding range in size from five people to more than 500. (One out of 5 of those submitted was from a firm of 16 employees or less.) Circulation varies from 200 to 3500, with 500 to 700 the norm. Sixty percent are published quarterly; the rest are either "irregular" or biannual. Length and format vary, but (in the opinion of the Coxe Letter) the most successful are two to four pages, one-color, 8½ x 11", with ample graphic and photographic material. Nearly 40% are self-mailers.

The best are brief, easily scanned and read, timely in information presented, soft-sell and educational in style. Unlike brochures, white space in newsletters is wasted space, and the most appropriate format is a compact one.

The firm logo and a standard mast head format are the principal means of establishing the identity of most publications. A third are named "ABC Newsletter" or "QRS News". Other typical names: Update, Outlook, Scope, Progress.

While some are topical in content, most feature the firm's recent work, including new commissions; projects underway; recently-completed or -dedicated buildings; and awards and other third-party recognition of work such as exhibitions and publications. Typically, three or four projects will be featured, with photographs and descriptive text; and in some cases other recent work is listed in a summary fashion. Successful headlines play down the firm and appeal to broader interests of the reader, e.g., "$1 Million Remodeling To Cut 10-Year Energy Costs by $2.5 Million," "New . . . Seen As Model For The Future," "Joint Venture To Design . . . Metro Centre," "John Deere Distribution Center Near Completion at Conyers."

Some newsletters include brief technical articles (e.g., the use of computer systems to save energy, value engineering, construction methods). "News" items, such as the announcement of public works authorization, state building code changes, and an energy bill update, are included in some.

People are also featured, more briefly, with announcements of promotions, awards, teaching positions, recent registration, seminar attendance, and so on. Often new employees are named, with short bios. Some newsletters "profile" a particular member of management or staff at more length in each issue. And the

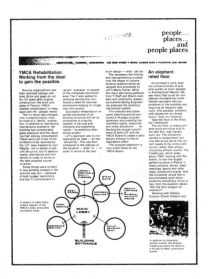

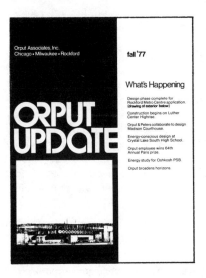

occasion of a firm's anniversary, office move or expansion, or addition of people and capabilities to a department is covered (or may become the lead story).

All of these elements are fairly common. What makes a newsletter good, however, is not the content so much as the writing and the "slant". The most successful newsletters are informative; projects are described in a way that is interesting to the reader. It's a "for-your-information" approach rather than a "look-how-great-we-are" approach; a "this is how the building works for the client/city" story rather than an "our project has x square feet and cost x dollars" story.

The intent of these newsletters is two-fold: to keep clients and friends informed about activities and capabilities of the firm, and to "educate" the reader about the process of arriving at a building's solution. They focus on the client of a specific project and his needs, his program, and how those needs were met. And they say it without using technical jargon.

Which brings us to the matter of who writes newsletters. Of those sampled, a fourth are written (in whole or part) by an outside consultant. Half use the talents of the PR/marketing staff, and almost a third use the technical staff at least in part. However, of those which are particularly successful, in the opinion of The Coxe Letter, more than 60% use outside writers.

Newsletters for staff and employees ("internal"). Distinct from external newsletters, the in-house variety are produced to enhance internal communication. They are inexpensively produced (often just typed and offset) on a fairly regular basis—mostly weekly or monthly. The names ("The Bumwad," "Addendum," etc.) express the humor and casual tone which characterize the content. Serious contents typically include: new commissions, status of ongoing work, upcoming interviews, departmental activities, office procedures, and morale-boosting messages from the president. On the lighter side, office parties are announced and postmortemed, classified ads are run, in-jokes are perpetuated, favorite recipes are shared, and people's doings are chronicled, with mutual back-patting all around.

To sum up: It is clear that firms investing in newsletters as a way of communicating to clients and friends are finding that it pays off (see Case Studies below). And for every firm that already publishes an external newsletter there's another firm seriously considering it. Newsletters can be, at least in the short-run, a mainstay of a firm's direct mail marketing program. There's evidence, too, that they are becoming a cost-effective way to supplement other marketing tools, especially that always-obsolescent necessity, the office brochure. And a newsletter can offset a lot of expensive contact time as a means of keeping in front of past clients. If you have at least the start of a good mailing list, and if the rest of your marketing program is running well, the newsletter may be a smart addition to your marketing efforts.

CASE STUDY #88: NEWSLETTER IS PART OF THE BUSINESS

James McGranahan Associates, a seven-person A/P practice in Tacoma, WA, publishes a single-sheet two-page newsletter called "Shop Talk" four times a year. (See illustration, Page 201.)

The lead article is topical; recent issues have featured fire and life safety, value engineering, and fast-tracking. Other items include "New Projects" and "Personnel Notes." To write it, principal Jim McGranahan drafts the lead item and talks it through with the firm's PR consultant; several edits later they're ready to go to press; 250 are mailed to friends and past clients.

Costs are hard to itemize because the PR firm is on retainer for other work as well, but McGranahan estimates the PR expense at $150/month, plus printing.

Is it worth the cost? "I think it is," says McGranahan. "We've had an excellent response . . . It's part of the whole marketing program. We're not professionals working in a vacuum. We're businessmen too."

CASE STUDY #89: NEWSLETTER KEEPS PROSPECT ALIVE

Chrisman, Miller, Wallace, Inc. of Lexington, KY also uses an outside consultant to help with its quarterly newsletter. "The intent," says partner Don Wallace, "is continuous name exposure."

Since publication started 1½ years ago, the 36-person firm's mailing list has grown from 500 to 1500 names. The audience includes past clients, local and state government officials, Chambers of Commerce, friends, and prospects.

Storylines are developed by principal Chenault Woodford and marketing coordinator Molly Sweezy. The PR consultant also helps decide content and drafts the four-page newsletter. Articles focus on specific CMW projects, but the slant is on the client and his design needs rather than the firm. Other items include a brief listing of the firm's on-going work, and "Inside the Company" covering the more human aspects of employees' lives. Cost run about $2100 per issue, half going for the PR consultant, a third for in-house time, and the rest for printing and mailing (it's a self-mailer).

"People have gone out of their way to say how much they enjoy the articles," says Wallace. "Occasionally we get a note of thanks from a client for an article about their project."

Wallace tells of an extra bonus: "We interviewed for an industrial feasibility study. We didn't get it, but as part of our follow-up we put the client on our mailing list for the newsletter. The client called me and said he appreciated it. Later he came back, invited us to interview for the ($1.25M) building itself, and we got the job." Says Wallace, "The newsletter has exceeded my personal expectations as a supportive element in the marketing program. It's another piece of a continuous marketing effort."

CASE STUDY #90: NEWSLETTER AMPLIFIES FIRM'S CREDENTIALS

The New York City A/E/P firm of Haines Lundberg Waehler has published a four-page newsletter on a "flexible quarterly basis" for four years.

"What makes ours distinctive," says Joan Capelin, HLW's Director of Public Relations, "is that we aren't selling a specific building, but rather a building type. I may refer to a project," says Capelin, who writes the newsletter, "but I'm not focusing on it. We emphasize problems that are common to many clients, but that the solutions are unique." The issue illustrated is on manufacturing facilities. Upcoming issues will feature energy opportunities, computer facilities, and medical facilities.

"The style is telegraphic," she continues. "The most important items are on the first and last pages."

The mailing list includes a base list of 850 plus the 300 employees (sent to

their homes) and hand-selected names from the firm's press list of 150. It's mailed flat in an envelope with conventional stamps, first class.

Printing and paper costs for 3000 run about $1200. Indirect costs are a week of a graphic designer's time, about 7 days for Joan's editorial time, and about 2 days of other administrative time. Mailing expenses are about $700 per issue.

The immediate impact of the newsletter is "subtle," notes Capelin. But the firm has designed the newsletter to serve a variety of purposes. While some 1400 may be mailed, 2000–3000 (depending on the topic) are printed. They're used as interview hand-outs, as part of qualifications packages, and as responses to requests for information on the firm. Says Capelin, "The newsletter has become a supplementary brochure for us."

IV

Making It
Happen

Many quite successful architects and engineers have no organized approach to influence the work they do. These firms simply respond to opportunities that come along. Such firms can become highly skilled at developing response strategies and making presentations to close the jobs, but in a marketing sense they are merely reactive—subject to the whims of the marketplace that seeks them. Their fate is out of their hands if a market dries up or a stronger firm comes along to compete.

This need not be so. Design firms can take control over the nature, volume, and location of the work they do. The following chapters address how to make this happen.

20

THE MARKETING PLAN

"The primary marketing problem is how to get ahead of the game. The way to do it is to begin with the plan instead of the problem."

Consultant John Coyne

The function of the marketing plan is to take the future of the firm out of the hands of fate and give some direction to what work the firm will do and for whom it will be done. This requires that a firm decide what it wants to be good at, select those markets where it can be most effective, and then go out and pursue the clients for whom it wants to work. This is the essence of planned marketing, as different from reactive selling.

To have a marketing plan one must first have marketing goals—for volume, type, quality and geography of work, etc. This is not difficult provided the firm itself knows where it wants to go. But for many architectural engineering organizations this is easier said than done. The process of setting overall firm goals can surface larger personal and management issues which, unless resolved, will stand in the way of successful marketing. Dealing with these aspects of overall goal setting is a separate management issue and is the subject of many separate books.* For marketing planning purposes, it must be assumed here that prior decisions have been made about what the firm wants to be. With this established, it is then possible to go about determining the firm's marketing goals.

SETTING MARKETING GOALS

The marketing goals, of course, are both long range and short range. Plans to enter new markets, add different services, shift to larger (or smaller) clients, etc. require strategies that may extend over two to five years. Shorter range, it is necessary to decide where the work will come from that will keep the firm busy in six months to a year.

For annual marketing planning purposes it is often best to begin with an analysis of the firm's previous year workload to establish:

1. *Carryover.* The amount of annual gross fees that come from work in process carried over from the previous year. For most firms this averages 30% to 50% of annual billings.

2. *Repeat and referral.* The amount of new work billed during the year that came as repeat work from former clients or by direct referral requiring little or no marketing effort. This category tends to be the largest source of new work in most firms, averaging 25% to 50% of annual business.

3. *Marketed New Business.* The work performed during the year that was obtained because the firm went out and sought it. This category tends to provide the smallest share of each year's billings—averaging 10% to 30%—but is, of course, the major generator of carryover, repeat, and referral work for future years.

When this analysis is expressed graphically it can look like Fig. 20-1A.

The next step is to analyze each of these categories by market (or client) and try to get a handle on where the work has been coming from. This is expressed in Fig. 20-1B.

If each of the markets in which the firm expects to be active is then projected, the composite of these individual market plans can produce a goal projection like Fig. 20-1C.

---

*One, by the author, is *Managing Architectural and Engineering Practice,* John Wiley, 1980.

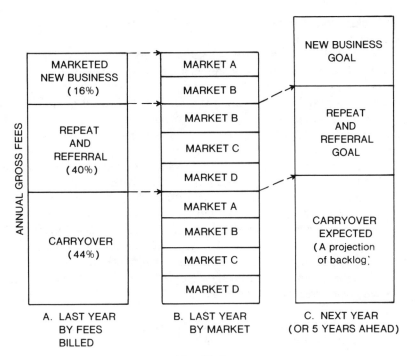

Fig. 20-1.

INDIVIDUAL MARKET PLANS

There are almost as many different approaches to constructing a marketing plan as there are different design firms. Some marketing directors prefer to begin with market research data that projects the external environment, and then plan how to respond. Others like to start with an assessment of internal strengths and then seek markets where those strengths are viable. Still others see marketing planning in terms of projected activity (e.g., number of contacts = number of leads = number of jobs), and base their plan around the effort needed to achieve objectives. All approaches are valid, and most marketing plans combine some parts of all three.

In practice, however, the marketing in most engineering and architectural firms is done by principal professionals and a marketing plan can be accomplished only if those people are organized and committed to carry out the essential tasks. It is therefore the author's belief that the place to begin marketing planning is with individual market plans which are acceptable to individuals in the firm who are committed to carry them out. These individual market plans then become the building blocks upon which a firm-wide marketing plan can be built.

A very simplified matrix for a one-market marketing plan is illustrated in Fig. 20-2. The first step in applying this matrix is to define the market being planned. This can be a specific project type for a building-type specialist, a territory for a branch office manager, or a selected list of past clients for someone who will concentrate on marketing repeat work. The important point is to define the market clearly enough so a contact list can be prepared.

The steps which follow, e.g. bird-dogging, courting, and pursuing the resulting leads, have all been explained earlier. What is important now, in order to have a valid plan, is to quantify the effort to be applied to these steps and the results expected. A plan which does not project a yield within a defined timetable is not a plan. Some approaches to measuring marketing activity are discussed in Chapter 24. What is important to understand here is the framework within which an individual market plan can be developed.

A complete one-market plan can be typically written in a one or two-page narrative having the following elements:

Market. A definition of the type of clients/projects to be pursued and the geographic territory to be covered.

Message. What the firm will sell to those prospects.

Contact List. The names of past, present, and/or potential clients identified as being in that market.

Marketing Method. How the firm will seek opportunities in that market. What will be the systems for lead finding, courting prospects, and presenting the firm's qualifications.

Marketing Tools. What sales tools, promotional mailings, publicity, advertising, etc. are available or needed to support the effort.

Marketing Organization. Who will be responsible for the firm's effort in the market; how much time they will apply; what budget will be allocated for direct expenses.

Market Goal. What yardstick will be used to measure progress (e.g., number of contacts made, jobs won, fees actually sold, or what?) and what results are expected in the period of the plan.

A market plan statement along these lines can be prepared for a market as small as a half dozen individual clients which a single practitioner (in a small or large firm) wants to cultivate over a year ahead; or for a specialized project type which will be the focus of an entire division in a large orga-

MODEL FOR A ONE-MARKET MARKETING PLAN			
	ACTION	GOAL / TIME	UNIT OF MEASURE
Step I	LIST BUILDING Task: Research the Market. Find sources to list "suspects".	By (date), complete list, in priority order for contact.	List to be at least (x) names to contact.
Step II	BIRD-DOGING Task: Contact the Suspects.	Identify by (date) those on list who would consider your firm if they had work.	(x) to (y) names worthy of continuing contact.
Step III	COURTING Task: Get acquainted. Keep in touch.	Convey firm's message. Be on speaking terms when live opportunities arise.	(x) visits / contacts over (y) time with each prospect.
Step IV	LIVE LEAD STRATEGY Task: Pursue opportunities that come from contacts.	Get the job.	Percentage of success.

Fig. 20-3.

nization. If the firm is new at planning, the projections will obviously be rough approximations at first. But once the plan is committed to paper, and monitored over time, experience will provide a basis for surprisingly accurate projections thereafter.

COMPOSITE MARKETING PLAN

Figure 20-3 illustrates a manner of consolidating individual market plans into an overall marketing plan for a firm. The matrix has been filled in for a hypothetical firm to illustrate some of the different considerations that might apply in a typical plan.

Note first that the strategy for the hospital market, for example, where the firm is top-heavy, is to rely on repeat work from existing clients, and direct referrals, to find leads for new work. While this market will account for 50% of the firm's work for the coming year, it will consume only 30% of the marketing effort/budget because the emphasis of this plan is on diversification.

In the industrial market, on the other hand, the plan calls for aggressive bird-dogging and prospecting to find leads, and a heavy PR effort to build the firm's identity in that market. Thus, the industrial market plan consumes 40% of the total budget, although it is only expected to yield 15% of the work for this period.

Of special importance is the market category headed "Targets of Opportunity." Virtually all firms are reactive part of the time, and gain a substantial share of work from opportunities which cannot be derived from targets of opportunity. This does not mean that the category cannot be considered in the marketing plan. In practice the amount of work that comes in through this source can often be quantified by looking over past records. The objective of the marketing plan is not to ignore targets of opportunity, but to plan around them so some of the firm's work is under more direct control.

The long range goals at the far right of the matrix in Fig. 20-3 are the key planning element. How much of the firm's work will come from each of its markets, now and in the future? Design firm markets have distinct life cycles, and any firm striving for stability of workload will want to balance its mix of work so that it can ride out the ups and downs of individual markets. (An exception would be a firm of specialists in one exclusive field, such as earthquake engineers. For such a firm, the alternative to diversification might be geographic dispersal, with branch locations chosen so that economic ups and downs might balance. In such a case, the markets might simply be defined by the different geographic areas where the firm plans to be active.)

To repeat what it is hoped is by now obvious: It is possible to construct a marketing plan for any size or nature of practice. The process may appear

MARKETS	NO. OF PRESENT CLIENTS (if any)	NO. OF POTENTIAL CLIENTS on list to be contacted	MESSAGE	MARKETING METHOD (e.g., prospecting, referral, community work, etc.)	SALES TOOLS (Method and Budget)	ASSIGNED TO:	ASSIGNED EFFORT (hours)	ASSIGNED DIRECT EXPENSE ($)	% OF TOTAL EFFORT	% OF TOTAL BUDGET	GOAL (Number of clients or $ volume)	% OF PRESENT WORKLOAD	PRIORITY	% OF FUTURE WORKLOAD
Hospitals	21	—	Track Record	Referral	Use Present Brochure	AB	4 hr/wk	$4000	10%	10%	6 new / $500,000	50%	3	30%
Municipal	5	3	Track Record	Repeat & Prospecting	Use Present Brochure	CA	8 hr/wk	$1000	20%	10%	$150,000	15%	4	15%
Targets of Opportunity	80	—	We are Local	Community Work	Newsletter	JB	12 hr/wk	$10,000	30%	50%	$200,000	20%	2	20%
Industrial	1	50	New Dept./ Low Cost	Prospect	New Brochure	BC	16 hr/wk	$15,000	40%	30%	2 new / $150,000	15%	1	35%

Fig. 20-3.

initially to be inexact, but firms that have tried it report surprisingly specific results. There is in fact a body of behavioral science which believes that most of what people aspire to, they can accomplish. Since architecture and engineering is so largely a people business, it is not surprising that firms which make marketing plans find that they are successful in making them happen.

from: **the Coxe Letter**

CASE STUDY #31: CONSULTING ENGINEER'S MARKETING PLAN IS TO HELP ARCHITECTS

Traditionally, most consulting engineers have allowed the architect to get the client, and have limited their own marketing efforts to cultivating different architects. An exception is the seven-year old mechanical electrical consulting firm of John F. Steffen & Associates, St. Louis, which has been taking the initiative to help architects get work with mutually beneficial results.

The program was organized a year and a half ago when Steffen decided it was time for his firm to do more than just local St. Louis work and he engaged Bill Flaig as director of marketing to help expand the practice. Flaig, a college english major and pilot who had learned marketing while operating a local airline, surveyed the situation and concluded that "most architectural firms desired help in managing marketing sales efforts." He concluded that if the Steffen organization could help architects with their marketing, the engineers would benefit too.

As a first step Flaig went to a half dozen other midwest cities and identified one or two architectural firms that are not in competition and that were willing to associate with Steffen on projects. Then he began assisting these architects with their marketing. Examples:

- In two architectual firms he has conducted in-house workshops to help the firms set up marketing programs including identification of markets, determining how to generate prospects, and how to assign responsibilities.
- In other cases he has also helped architects find services other than design (e.g., medical facilities programming) that can be offered as an opener to lead to later design work.
- Flaig maintains his own lead-locating system through a clipping service and telephone calling because, "There's a lot of activity that doesn't come to architects attention and many times we can provide them with viable prospects."
- Wherever possible Flaig arranges for engineers to attend interviews with the architects in order to demonstrate their strengths in computerized cost control and scheduling.
- In one case Flaig helped put two of his architect clients into a joint venture that is now actively seeking federal work in their area.

The program is producing results for the architects—and for the engineers. In the year since the program got actively underway the Steffen firm has taken on projects in five cities in addition to St. Louis.

CASE STUDY #96: 'POSITIONING' BRINGS RESULTS

The Leonard Parker Associates is a 12-person architectural firm in Minneapolis with a strong design orientation. Three years ago the firm was in a slump, and efforts to get work weren't paying off. "We sat down and talked about it," recalls Director of Business Development Sandra Johnson.

At that meeting were President Leonard Parker, the four vice presidents, and Johnson. "Someone said that our image was 'too expensive'—it was the first time that had been vocalized," notes Johnson.

"Parker, who had been with Saarinen's office, then said he had worked for thirty years for quality design and he didn't want to give up that image. We realized that our buildings look more expensive than they are. We didn't want to lose the high-quality image, we just wanted to lose the high-cost image."

As a result of this meeting, the firm took deliberate steps to reaffirm its design image. The vice presidents began to concentrate their marketing efforts on design-oriented clients. (Most of the firms' work is corporate and higher education.) "We're going after the same clients but we're doing it differently," says Johnson. "We choose our projects more carefully, and look for a good match between what they want and who we are."

The firm is also making a conscious effort to communicate that high quality design doesn't cost more. "We make sure that a prospect knows about the smaller and less expensive projects we've done."

"We also started to stress cost in our slide presentations," notes Johnson, "to show that most of our work has come in on or under budget." She's now making some slides on square footage costs for further documentation of the new message. One project that will surely be included is the Law School of the University of Minnesota, which came in $700,000 under budget, cost $42.33/s.f. and has won five national and state design awards since January.

"We made a conscious 'positioning' decision," says Johnson in summing it up. "We reconfirmed our image, and it's successful because we're really working at it."

As evidence, the firm has just been chosen over four top national design firms as architect for the Hubert Humphrey Institute.

CASE STUDY #99: HOW TO SHIFT A MARKET MIX

Three years ago, virtually all commissions obtained by SKBB, Inc. of Coral Gables, Fla. were from the private sector. Fluctuations in the demand for design services meant that the now-20-person firm would have to expand its marketing plan to maintain the growth from a small firm to a medium operation.

An analysis indicated that the firm's talent and experience—which consisted largely of hospitals, schools, and major residences—was adaptable to public work. The decision to attack the public sector also coincided with the enactment of a "competitive negotiation" regulation which required all government agencies in Florida to advertise their search for professional services. The regulation eliminated the "sometimes horrendously political" aspects of cultivating public work, in the words of Robert J. Boerema, currently SKBB's director of marketing and president of the state's AIA when the rule was passed.

"You still have to become known," noted Boerema, "but you respond to specific notices for projects of defined scope. You can therefore be more selective, which is especially important in an unproven area."

The firm's new marketing plan called for each of the seven partners to accept responsibility to follow the potentials afforded by public prospects at one level or another. The assignments were based on both expertise and civic involvement. For example, Glenn Buff, SKBB's design partner, is often involved in presenting to public boards and civic meetings, so he was assigned to follow the agencies involved in the new public transit system.

The public area is unproven to SKBB no longer. Since adopting the marketing plan to go after public work, the firm has been commissioned to do a phased $9 million project for Florida International University; a federally-commissioned post office for downtown Miami; a county hospital; a $10 million airport concourse; and joint ventures to design a station for Dade County's new rapid transit system and the $24 million first phase of the Downtown People Mover.

21

MANAGING THE MARKETING PROGRAM

"Does the principal have to bring in all the work? . . .
Or, is the principal the one who brings in the work?"
Asked at a marketing seminar

Before any marketing plan can be implemented, it is necessary to identify those who will accept principal marketing roles in the firm.

In most design firms, the key players are a given group of professionals who have already chosen to work together. For such firms, marketing planning must begin with roles that are realistic and practical for the individuals involved. How the marketing functions are divided among the players in any individual design practice is a combination of basic philosophy of the firm, market acceptance, and personal aptitude of the principals.

The point of this chapter is to focus attention on the importance of deciding where responsibility and authority for managing the marketing effort will be structured. Many professional firms, in an effort to "solve" a marketing problem will engage or appoint a strong marketing director in order to create or give the firm a marketing program. This approach will succeed only if the marketing director has the authority to make the decisions necessary to implement a marketing plan—decisions like what kind of services the firm will offer, what kind and quality of work the firm will do, and who will carry out the marketing assignments. In most firms these are not decisions which the principals will allow someone else to make for them. The principals of most firms would be delighted to find a magician who can produce work for them, and will give that wonder-worker free rein, provided it requires no change in what the principals are or must do.

This is the essential dilemma in managing the marketing plan in an architectural or engineering firm. Marketing is just not an area where unilateral authority is very effective. Successful marketing begins by dealing with the heart and soul of the firm. Until certain structural decisions are addressed, there will be no effective plan.

It is for this reason more than any other that many architects and engineers do not have organized marketing programs—they simply have not resolved the management issues and conflicts that come up as soon as a marketing plan is discussed. In these firms no hired marketing director can do it for them, and no marketing consultant can create a successful plan that avoids the issues. It is best for such firms to face the facts that the personal styles and goals of the principals are not in harmony with an organized marketing approach and, instead, accept the consequence of relying on reactive selling to obtain work for the firm. There is nothing wrong with not having a marketing plan, if the decision is made as a matter of choice. A large majority of professional firms will probably always make this choice.

The reason for this gets into the essence of the professional relationship. If the client and the engineer or architect are to have a serious agency relationship, nine times out of ten that relationship will be cemented prior to or at the sale. In fact, closing the sale is usually the first act of agency between the two. Thus, anything which gets between the client and the agent prior to the sale tends to color the total relationship thereafter, and often turns the professional services into a product or commodity.

It is also true, more often than not, that: "Those who can get the work, run the firm." It is like a law of nature that those professionals who can bring in work either become principals of the firms they are in, or soon go into practice for themselves. Thus, the role of the "closer" most frequently will be held by a principal. (There are some notable exceptions, especially in large engineering firms, where getting the job is the responsibility of hired closers who work in a marketing department, but these organizations tend to be selling specialized services based on a reputation created by a previous generation of principals. The ability of the "hired closer" marketing organizations to function consistently over time remains unproven.)

A marketing organization composed of closers who close the jobs and then hand off the work to doers who do the projects is quite different from an organization where the closer is also a doer, and remains involved after the sale to manage the work. In the 1960s and early 70s, many marketing organizations built around charismatic closers and separate doer/project managers were very successful. But in recent years clients have demonstrated a clear preference for marketing organizations composed of closer-doers, where the sale is made by the professional who assures the client they will be personally involved, to a credible degree, during the execution of the project.

The significance of this in marketing management lies in how control is structured. Since the closers are presumably principals of the firm, they are not likely to delegate to a marketing director the authority to tell them what to do and when to do it. In function (if not by title) the closers are themselves the marketing directors, and the marketing organization in such firms is derived directly from how the closers and the doers are structured. This is illustrated by the two charts in Fig. 21-1, which apply a marketing function to a typical organization structure.

In M-1, a directed marketing program, the assumption is that a strong marketing organization has responsibility for finding and closing the clients and delivering them to the project management professionals who may appear at interviews but are otherwise involved as little as possible prior to the sale.

In M-2, a facilitated marketing program, the assumption is that the leading client management professionals in the firm will have primary responsibility for selling and finding clients. The marketing organization is there to support and coordinate to whatever extent the individual professionals may require, but the ultimate accountability for work under contract remains in professional hands.

The choice of marketing organization structure should be made according to how the closers want to operate.

A firm where the principals who are 'close' are able to hand-off the client and the "doing" to project managers will find a directed structure (M-1) appropriate. On the other hand, a firm where the principals are closer/doers and stay heavily involved with the client after the sale will generally find the facilitated marketing structure (M-2) more appropriate, because the strength of the closer/doer is derived from the clear free-of-handoff link between the client and the professional. In an M-2 structure the closer/doer may use marketing staff for lead finding and strategy but most of the courting will be done by the closer.

The approach to managing the overall marketing program will also be

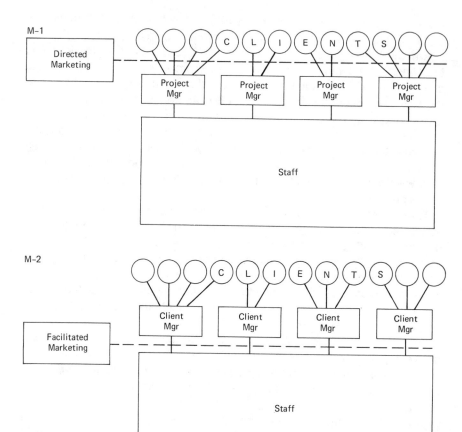

Fig. 21-1.

significantly different according to which structure is adopted. The marketing director in an M-1 directed organization will most frequently be a principal professional in the firm who has the controlling power to make the closing decisions. This director may employ a marketing manager to implement those decisions, but is not likely to let the hired manager make policy. In an M-2 facilitated organization, on the other hand, the marketing manager is generally coequal with the closer/doers—a team player with a collaborative voice in policy. On this basis the M-2 marketing manager can be either one of the professionals or a trained marketing person.

Both modes of marketing organizations are being tried in design firms, but with notably different emphasis. A decade ago, when a significant number of design firms had just installed formal marketing organizations, the ratio was approximately 50% directed and 50% facilitated. At this writing, the proportion is more like 90% facilitated and 10% directed. The firms that abandoned the directed approach did so largely because they found in practice that it did not lead to the best professional relationships, and in many cases it was more costly to sustain. Those that are continuing with the directed mode are chiefly firms where the service very often is more "product" oriented—such as specialists in one project type—and where there is less of an agency relationship with the client.

Thus, a very important part of the marketing planning process is the decision within the firm as to how the program will be structured and managed. Doing the marketing requires the same functions in either mode, but managing it so it all actually happens is the key element of a successful marketing plan.

---

from: **the Coxe Letter**

CASE STUDY #98: IN-HOUSE WORKSHOPS TRAIN MARKETING TEAM

About a year ago, Hammel Green and Abrahamson, a 115-person firm of architects and engineers in St. Paul, noticed they were coming in second in too many interviews. As a result, they started to focus on ways to train the 21-member marketing team. (The marketing team includes the vice presidents in charge of 7 divisions, 3 founding partners, 9 technical people, and the 3-person marketing staff.)

Joanna Baymiller, the firm's public relations specialist, talked with two consultants about having them conduct a training program. "We became convinced they would deliver an 'event,' and we wanted a 'process,'" Baymiller reports. Working under the direction of Marketing Director George Riches, Baymiller decided to create the training program in-house.

She sent out a questionnaire to the marketing team which asked what they most wanted help with. Their first priority was using the telephone and phone cold-calling techniques; the second, writing; and the third, interviewing techniques.

Baymiller then designed the first training session on "Phone Power." To prepare, she read some books and interviewed people at the phone company, which offers training programs. The 2-hour session was attended by about a dozen people. Working in groups of 4, two participants role-played prospect-architect phone conversations (Baymiller gave them a list of "situations" from which to choose). In each group two people listened and took notes. Tape recordings were played back for the whole group, and the "listeners" shared their observations. Baymiller then passed out some information, gleaned from a book on successful phone-call techniques. The role-playing was then repeated. "Everyone did much better on the second round," says Baymiller. "I had no magic wand or bagful of tricks. They learned by hearing themselves and getting feedback from each other."

Six weeks later a second session was held, this time on writing. Baymiller defined "situations" that would need a letter response. Each attendee wrote a letter, shared it in small groups, and the entire group selected and then critiqued the "best." She also reviewed basic rules of writing and distributed handouts. One of the 18 people attending was a founding principal, who said afterwards, "I had my doubts about this but it turned out to be worthwhile."

22

STAFFING THE MARKETING EFFORT

"Marketing people must be self starters who don't have to have someone tell them what to do every day."

Marketer R. Paul Wood

"A good marketing manager can relieve you of 99% of the headaches you have in marketing your services. But the marketer is never going to replace the licensed professional. I see my job is to bring the client and the registered professional together in an atmosphere that is conducive to creating opportunities for good design."

Marketer Robert P. Boone, Jr.

"When you get him or her on board your work has only begun. . . . Your real effort must go into training, orientation and the support necessary to help the marketer become effective."

Manager Louis L. Marines

As engineers and architects have accepted the principles of marketing, there has been increasing experimentation with ways to delegate some of the marketing effort to persons who can specialize in marketing tasks. This is reasonable, because most design professionals did not go to architectural or engineering school in order to spend the rest of their careers ringing doorbells. And since clients are showing a clear preference for working with professionals who are both closers and doers (See Chapter 21), it is important for those with these skills to free themselves from as much of the mechanics of marketing as possible in order to serve the clients they win.

225

In this climate, design firms are finding that it is possible to delegate to persons with specialized marketing roles from 30% to 70% of the hours required to carry out an effective marketing effort. This chapter will try to define those roles and address the experience firms are having in finding qualified persons to fill them.

As used in this chapter, the term "marketer" refers to anyone holding an essentially full-time staff role devoted to marketing tasks, as distinct from those design professionals who may market with one hand while carrying on their professional practice with the other. An increasing number of persons trained in the design professions are setting aside their technical skills in order to devote their careers to marketing roles. These persons are nevertheless "marketers" as defined in this chapter.

The significance of delegating as much as possible of the marketing tasks lies in the ratio between the number of closers in an organization and the maximum size at which that firm can efficiently operate. With marketing costs in architectural and engineering firms averaging 4–8% of gross fees (see Chapter 23), it is clear that one closer in an organization much larger than 15 people will have little time for doing—or anything else—unless the closer is able to delegate some of the marketing functions. This, in fact, appears to be the experience in practice. The single principal of a small firm can manage both to get and run the work up to a staff size of 5 to 10, with no marketing help except possibly the assistance of a secretary taking care of the paperwork and sales tools. A firm of two or three similar principals who market can grow to 15 to 30 staff in the same manner. Beyond this size, however, the effort to feed work to the firm becomes more than a part-time job. When this happens, one of two directions is possible: One of the principals can become a full-time marketer and turn over some of the professional doing to others; or some full time staff assistants can take over the marketing support roles.

In order to decide what to delegate it is first necessary to define the different marketing functions in the firm and to evaluate the time devoted to each of them. Surveys of marketing people in a variety of different size architectural and engineering firms indicate the average division of time among the major tasks is:

5–10% is *marketing management* including market research, planning, and day-to-day decision making.

25–35% is *lead finding* including both cold calls, following up with past contacts, and qualifying targets of opportunity to determine their validity as leads.

15–25% is *courting,* including both romancing of live leads and long-term courting of potential and past clients.

20–30% is *coordination* and pulling together of statements of qualifications, proposals, presentation plans; generation and maintenance of sales tools; and general internal support of the marketing process.

15–20% is *presentations and closing.*

In the light of this it is clear that the policy and direct selling functions which require the principal design professional aggregate only 30–50% of the total effort in a marketing program. Of course, in a wholly reactive selling program, where there is no planning and little or no lead finding or courting, the proportion of principal time required is much higher. In these reactive programs, about the only task that can be delegated is the coordination, which may aggregate 15–20% of that marketing effort. The breakdown of roles which can be delegated in a marketing program has a great deal to do with the nature of the firm, the scope of its marketing program, and the share of the marketing tasks which the principal professionals choose to perform.

In general, what can be delegated falls into one of three categories, and these are becoming identifiable disciplines within the world of design firm marketers. At this writing, perhaps 20% of all architectural firms and 10–15% of all engineering firms have formalized marketing organizations where there is some delegation of tasks. Of these, the employment of specialized marketers divides into the three categories approximately as follows:

About 10–15% of the firms have *marketing managers/directors* who are responsible for pulling together the marketing plan with the firm's management, and then leading the marketing organization that will carry it out. Where the role exists it is generally a senior staff function reporting directly to a marketing or managing principal or an executive committee. Salary levels for marketing managers/directors were generally in the range of $30,000 to $50,000 in 1982 dollars.

Some 20–25% of firms, including about half of those which also employ a marketing director, are employing *marketing representatives* (birddogs) whose primary function is to seek and pursue leads for one or more closers. Many firms which use this role employ several marketing representatives, with each concentrating on a different market. This role was paying $25,000 to $40,000 in 1982 dollars.

About 95% of the firms that delegate any marketing have a designated *marketing coordinator* to pull together the internal marketing resources, help prepare questionnaires and statements of qualifications, develop sales tools, and provide general support for the marketing effort. The marketing coordinator usually reports to the principal in charge of marketing or the marketing manager / director if there is one in the firm. Marketing coordinator salaries were in the $15,000 to $35,000 range in 1982 dollars.

As more firms have created these roles and more and more marketers have tried to fill them, a pattern of job descriptions and successful qualifications has begun to emerge that can provide some points of reference for firms considering delegating some of their marketing tasks. Nevertheless, it is important to recognize that all the positions are relatively new and it is not yet possible to treat them as formal disciplines with established qualifications and performance standards. Thus, firms which are considering filling the positions must be prepared to experiment and learn together with those hired or promoted to fill the roles.

Following are discussions of what is currently known about the criteria for filling each of the roles.

THE MARKETING MANAGER / DIRECTOR

The job description and duties of a marketing manager and a marketing director are essentially the same, but there is a world of difference in the personal style called for by the two positions.

The fundamental role of the marketing manager / director is to see that the firm has a comprehensive marketing plan and then organize the marketing activities of the firm to implement it. A detailed job description for the role is included in the Appendix.

The appropriate approach to accomplishing this function will be substantially different depending on where control and direction of the marketing program are vested. As discussed in the previous chapter, in most design firms the principals do not—and very probably should not—ever delegate to staff the authority to control or direct much of their destiny. Thus, a staff marketing director would be inappropriate in such firms because the title implies more authority than these firms will delegate to staff level.

A staff marketing director position, by definition, is chiefly valid in firms where large segments of policy are staff generated, such as in some of the larger engineering firms and those few design firms which are externally owned by investors rather than owned by their principals. In almost all other firms, the marketing manager title is more appropriate. The role of the marketing manager in control and direction is to bring the appropriate issues and considerations to the attention of the principals and to facilitate the decisions necessary to establish marketing policy—but not to direct those decisions.

It is on this narrow but essential point of definition that successful filing of a staff marketing manager / director position rests. A large part of human nature will seek to control events whenever the opportunity seems present. In many design firms the marketing manager / director fills a relative vacuum. If that person's personal style is control oriented, they will try to fill the vacuum by deciding such matters as where to pursue work, who should go after certain projects, etc. If the principals resist this sort of authoritative approach, the relationship usually deteriorates and, in time, the marketing director moves on in frustration while the professionals in the firm begin to have doubts about whether they can ever staff the marketing function.

The personal style that is more successful in the marketing manager position is that of a coach or team player who enjoys and is skilled at motivating others to do their best, while not personally requiring the last word on key decisions. This is a much more difficult style to perfect, and since it goes against much of the authoritative training that is instilled in our culture, it is often difficult to find candidates with this skill.

The first step to successfully differentiating a marketing manager from a marketing director is to be clear from the outset about the difference in style that is expected and to make sure candidates being considered understand it.

Once this difference is clear, the chief qualification required to fill a marketing manager / director position is an appreciation for professional services as distinct from products or commodities. Few successful marketers for design firms have come from industrial or product marketing backgrounds. There appears to be too great a gap in philosophy—or perhaps in salary levels—for many to make the conversion. Thus the better "outside" candidates for staff marketing manager / director positions are coming from the softer disciplines such as banking, city management, journalism, etc.

For all the reasons cited above and in Chapter 21 it is becoming apparent that the best candidates for marketing manager / director positions in design firms are persons with prior training and / or experience in the professional disciplines. Some are registered professionals who have redirected their careers; others have some education in engineering or architecture and have been attracted to marketing roles without undertaking the technical apprenticeship necessary for registration. Many of the most successful marketing managers were trained as professional planners. It is likely this trend will continue, for marketing policy and management is too close to the essence of professional design firms for the role to be filled by persons far removed from the field.

THE MARKETING REPRESENTATIVE

In staffing terms, it is practical to delegate about 70% of all lead finding to a professional subordinate or hired marketing representative. In doing so, however, it is important to understand the behavioral requirements of the bird dog role, and to be realistic about the expectations placed on persons who have the aptitude to do it. First of all, the ability day after day to go out and knock on strange doors to feret out information requires a high tolerance for rejection. Furthermore, since the bird dog's job is only to find the lead, and then turn it over to a closer to make—or lose—the sale, it is a "no win" role in which bird dogs seldom can take singular satisfaction for the success of their efforts. The most distressing events in the life of bird dogs are the days they know they should not go to interviews (where they might look like salespersons), even though they have been almost solely responsible for the firm getting to the interview in the first place. To have the closer then blow the job at the interview because the closer didn't follow the strategy that the bird dog knew was correct can be devastating.

This combination of behavior (high tolerance for rejection plus no need to 'win') makes good bird dogs very hard to find and harder to keep. Good bird dogs, when found, need to be given a lot of latitude around how they operate, and they should not be required to perform detail tasks, such as writing proposals or filing complex reports. Such disciplines rarely are compatible with the skills that make the bird dog role tolerable.

In practice, the ability to master these behavioral requirements seems to be a more important qualification for the role than any particular background. There are, however, some similar functions where people with compatible experience can be found, such as industrial development representatives of state and local governments or utilities, or marketing representatives of banks. One source that is often tried but is seldom successful is former employees of the project area to be bird-dogged, e.g., former government officials, former school or hospital superintendents, etc. While these people may perform other useful marketing functions, they rarely work out in the bird dog role because the requirements of their prior careers seldom include the behavioral skills essential to successful bird-dogging. Knowing their way around the field/industry from which they came is not the same thing as having a high tolerance for rejection and no need to "win."

Many of the most successful bird dogs experience burnout after three to five years of cold calling. This is consistent with the experience of industrial and insurance sales, where successful salespeople will spend only a few years building a territory of new contacts, and then rely chiefly on recalling these established contacts for their continuing business. The professional services bird dogs who must turn their established contacts over to others in the firm tend to run out of steam after a few years. The bird dogs that seem to last longest in the role are those covering markets (e.g., government agencies) where they do not have to hand off all good relationships.

For all these reasons, many firms are finding that bird-dogging is not a realistic career position. Instead, these design firms are either making it a part-time function of junior professionals on the way up, or are encouraging their closers or marketing coordinators to do some bird-dogging in addition to their primary roles. There is reason to expect this trend to continue. This means that most firms will make bird-dogging a limited, part time function of the marketing manager/director, or of assigned marketing principals, or will open up new markets with a one-time crash bird-dogging effort by temporarily assigned professionals.

All of this, however, is not meant to discourage the creation of staff marketing representative positions in those markets and in those firms where it can be effective. In practice, how much of the lead-finding can be delegated—or not—is often a function of the individual market. Public agency and institutional markets lend themselves to delegated lead-finders because they operate in an open environment where information about up-coming work is publicly accessible. Lead-finding in private markets increases in difficulty according to the decision structure in the client organization. Large corporations and utilities will be relatively more open than solo entrepreneurs, such as developers. Thus, the engineer specializing in subdivision development will generally have to do his or her own lead finding and courting, while the architect who specializes in corrections facilities can efficiently employ a bird dog to find and qualify leads before getting the closer involved.

The bird dog role can be a very valuable part of the right kind of marketing organization. A sample job description for the position is included in the Appendix. The objective is to be certain in the first place that there is a proper role in the firm for a full-time marketing representative, and then to try to match that role with the behavioral skills that will enjoy and be effective at it.

THE MARKETING COORDINATOR

The marketing function most practical to delegate is internal coordination. The role of marketing coordinator in design firms has grown so rapidly in recent years that it has become a paraprofession of its own. The function can pay its way as a part-time activity in firms as small as a half-dozen people, and it is becoming a common full-time activity in firms of 15 to 20 persons and up.

The value of the marketing coordinator is that the functions performed by the role tend to be most efficient—and effective—if they are specialties of one person. These include compiling the detailed statements of qualifications and staff resumes required by so many RFP's; maintaining resource materials such as photographs, slides, job histories, brochures, etc. in accessible order; and keeping a log of all the leads being pursued by different principals and coordinating the marketing communication between them. A detailed job description for the marketing coordinator role is included in the Appendix.*

The marketing coordinator role is also the easiest of the staff marketing positions to fill because it appeals to a broad variety of backgrounds. Many of the best marketing coordinators are former executive secretaries or administrators in their firms and bring to the role an extensive knowledge of the firm's history, its people, and its projects. They are thereby readily equipped to know the facts necessary to put marketing packages together, and they know who to ask to get help filling gaps.

When hired from outside the firm many marketing coordinators have PR backgrounds because the role is best filled by persons with strong writing skills. It is important, however, not to let the role be confused with a PR/ communications position. The marketing coordinator who thrives on writing professional qualifications presentations and assembling answers to RFP's is not always the same person who can write and place product or institutional publicity. In addition to writing skills, a coordinator should be well organized, and able to work under pressure in an environment where priorities are constantly changing as new deadlines come up.

The combination of responsibilities and skills required of a marketing coordinator makes it an exciting position for younger people at an early stage in their careers. Many move on after two or three years to do other things, so firms are finding it best not to try to fill it as a career position. Rather, it is advisable to look on the marketing coordinator role as stepping stone for talented people, and keep an eye out for a reserve candidate at all times. (It is also essential that the manager of the marketing coordinator, whether a principal or a marketing manager / director, be sufficiently familiar with the role and its systems and procedures so that turnover does not totally disrupt the marketing process.)

COMPENSATING THE MARKETING STAFF

To date, the least refined aspect of marketing management in design firms are methods of motivating performance toward fulfillment of the marketing plan through incentive compensation.

In product sales organizations the traditional method has been to offer healthy commissions, concentrate management effort on training salespeople, and to measure performance in units sold.

The same carrot / stick methodology of sales management does not work for architectural and engineering service. There is no "product" that sales-

---

*An excellent monograph entitled "The Marketing Coordinator," by Janet Goodman, is sold by the Society for Marketing Professional Services, 1437 Powhatan, Alexandria, VA 22311.

people can be trained to sell over and over. Compensating marketing staff with commissions has not been found practical in design firms because of the team nature of the selling process. To reward a lead finder based on leads closed, for example, ignores the reality that the success of the sale is not in the lead finder's hands. Conversely, firms have been reluctant to pay commissions for leads found, regardless of whether they are closed. Compensation surveys among design firm marketers consistently show that 95% or more are employed on a straight salary basis, plus regular profit sharing or bonus. The great majority of those firms that have reported paying a commission at one time have discontinued the practice after a few years.

This does not mean that it is not possible to offer incentive rewards to marketers—professional or nonprofessional. The key is to find a way to relate the incentive to the activities which can be measured. As discussed above, the "sale" itself is only one of many elements of marketing activity which can be measured, and is generally the one element where it is most difficult to identify and measure one individual's contribution. At the same time, most firms have been reluctant to offer financial incentives for activity prior to the sale, relying instead on judgmental bonuses to compensate for outstanding performance. Much of the reason for this may be the lack of understanding of other ways to measure marketing performance. As the yardsticks outlined in Chapter 24 are put into practice, it is likely firms will find more creative ways to reward the members of the marketing team who achieve their goals.

INTEGRATING THE MARKETING ORGANIZATION

As the marketing roles that can be delegated become more distinct, there is a growing awareness that the way marketers filling these positions are being integrated into design firms still needs attention. The symptoms are a high turnover and mobility among marketers, and a continuing reluctance by design firm managements to give marketers a real place in their inner circle—especially any piece of ownership.

A case can be made that the most successful, long-term marketers in the design community are either design professionals per se, or are those "outsiders" who have most thoroughly grasped the special nature of professionalism that is different from business. It is also interesting to observe that the marketing coordinators as a group do not suffer the same symptoms as do so many of the senior level marketing managers/directors/representatives. It is possible that the coordinators see their role wholly as support for design professionals, whereas many upper level marketers keep wanting a voice in telling the design professionals where to go.

Design firm principals must come to terms with these issues if they are going to be successful in staffing much of the marketing function (as opposed to doing it all themselves). The entire marketing organization must be working toward the same professional goals as the principals. This means that the marketers and the design professionals must share the same values for the kind of clients the firm wants to serve and the kind of work the firm wants to produce. If the professionals themselves are not in agreement on these goals there is no way the marketing staff can resolve them. And if no one takes the time to educate new marketers to those goals, there is no way the marketing staff will understand them. Thus, the very best way to staff a marketing organization is to be very clear about the mission of the firm and then recruit and promote to marketing roles persons who share that mission. When such marketers are found, there should be no hesitation in allowing them into the inner circles of the firm's future.

from: **the Coxe Letter**

CASE STUDY #97: EVERY EMPLOYEE A MARKETER.

The principals of Chrisman, Miller, Wallace, Inc. (CMW) of Lexington, KY, wanted everyone on the 40-person staff to become involved in marketing. So, in 1976 all technical personnel began receiving copies of CMW's marketing program, and were encouraged to notify the firm's three-person marketing group of any prospective clients—however seemingly obscure—which might be pursued. Last year, nonmarketing employees suggested over 100 such leads.

The annual marketing program includes factors considered in developing the marketing strategy; a summary of services to be offered; the firm's organizational chart; geographic areas and types of clients targeted for intensive marketing; information concerning follow-up contacts with past clients; standards for accepting new work; forms and instructions for the submission of leads and topics of potential interest for CMW's newsletter or external publicity outlets; and various support materials, e.g., AIA's recommended compensation schedule.

All employees are issued business cards and, when appropriate, follow-up correspondence to prospective clients is sent out under the signature of the staff member turning up the lead. Thereafter, the technical employee receives copies of all correspondence between prospect and marketing group. Each employee suggesting a lead, irrespective of its ultimate feasibility, receives a note of thanks from CMW's director of client relations, Don Wallace.

Once a lead is deemed worthy of further pursuit, it is added to a list, with credit for the lead attributed to the deserving employee. The list is updated at monthly meetings of the entire staff during which about one third of the discussion is focused on marketing topics (projects won and lost, as well as strategies for selling new prospects). Each member of the staff is also asked annually to make a rough estimate of how much time he expects to spend on marketing-related activities.

Commissions cultivated through the program range from an addition to an industrial facility to the master plan for a church-supported housing community for the elderly. Besides such tangible benefits, the program has done much to diffuse the traditional alienation between the design and promotional staffs, says Molly Sweazy, CMW's marketing coordinator.

"If the technical staff feels any hostility toward the marketing staff, it's usually the result of a lack of understanding," she says. "The marketing program clears the air of a lot of the mystery which has always surrounded marketing."

23

MARKETING
BUDGETS

*"At least one hour of every 20 of principal time should
go toward getting the next job."*

Widely quoted axiom

Average marketing costs in design firms have been surveyed regularly for
more than a decade. These surveys reflected a gradual increase in budgets
until the late 1970's when marketing spending leveled off. At that time the
average architectural firm with a formal marketing plan was spending 5–7%
of gross fees on marketing, and an average engineering firm was reporting
marketing expenses in the range of 4–6% of gross fees. These averages
seem to apply quite consistently in firms having annual gross fees of
$400,000 (6–8 people) and up.

The leveling off appears to be a consequence of both what clients are requiring in the way of marketing effort during the selection process and what design professionals are willing to go through to get work. It is reasonable to assume these levels will prevail for the foreseeable future.

What is most significant about the surveys of marketing budgets is how the money is spent. With few exceptions, firms report that 70–80% of the marketing budget goes for personnel time, and only 20–30% for direct expenses such as brochures, travel, entertainment, etc.

Extending these averages means that for every 20 people in a design firm, the equivalent of one full-time person must be devoted to marketing: In a 10 person firm, at least half of one person is required. It is failure to accept this reality, more than anything else, that restrains the success of many talented architectural and engineering firms. Wherever the design professionals primarily enjoy the "doing," the temptation is to neglect marketing as long as there is work to do. However, with the typical lead time to develop significant work often six months to a year, such firms usually experience very erratic ups and downs in their workload.

Within these broad averages there are wide variations in how marketing expenses are budgeted and managed. Some firms maintain meticulous computer records of dozens of categories of marketing expenses and track by number the cost of pursuing every major lead. Others report only a bulk number made up of all time and direct expense charged to marketing and cannot even break out the difference between principal time and staff time. The degree of sophistication in tracking expenses seems to have little to do with success of marketing from one firm to another.

There are, however, a few guidelines that may be helpful for firms that want to establish some budgeting policies.

First, it is important to understand that marketing is an overhead function, and therefore marketing budgets always express personnel time at its raw costs rather than at billable rates.

It is worthwhile to budget the time of individuals with major responsibilities for marketing, and then be able to monitor whether the time is being spent. Thus, if 30% of Partner A's time is committed to marketing, the marketing manager should have a means of knowing if that time is actually being used.

There is a wide difference of opinion about whether regular accounting of the costs for individual job pursuits is useful. Once a commitment is made to go after a given project, most firms spend whatever they have to spend to pursue it, and knowing the exact amount later is not especially valuable. It is the total yield from the total marketing budget that is the real measure of marketing value. On the other hand, projecting in advance the probable cost of pursuing given projects can be very useful in making "go/no-go" decisions. Some firms establish policies that require approval from top management whenever the estimated marketing cost for a given prospect is more than, say 2% or 5% of the estimated fee. Once a "go" decision is made, however, tracking the actual cost of that pursuit is seldom worth the bookkeeping effort to produce the numbers.

Major long-term marketing expenses, such as producing a new brochure or opening a branch office can be amortized over their useful life rather than allowed to inflate one year's budget. Thus, a $30,000 brochure with a three-year life can legitimately be charged to the budget at $10,000 per year.

The real value of having a marketing budget—rather than a record of marketing expense—is the effort that is required to prepare it. This involves projecting in advance what marketing tasks will be done and what effort will be devoted to accomplishing them. That, of course, is what marketing planning is all about. Thus, the preparation of a marketing budget is an integral tool in helping to make marketing happen.

from: **the Coxe Letter**

CASE STUDY #16: A MARKETING BUDGET PAYS OFF

Before Architect Romeo Garcia established his practice in Corpus Christi, Texas, 2½ years ago, he had read AIA manuals that mentioned the importance of budgeting specific money for marketing. Thus, when he borrowed working capital to launch a one-man firm he included an allowance for "public relations." Thereafter, he set aside 10% of all collected fees for promotion activities.

The budget added up to only several thousand dollars the first year, but Garcia reports that he "made it a point to spend it." The principal expense was for travel to points where he thought there might be work (especially the regional offices of federal agencies); for lunches with prospective clients; and to support involvement in civic affairs.

Garcia's initial workload consisted of "a couple of residences, a church remodeling, and a small commercial office." After exploring available long-term opportunities he decided to concentrate his marketing in public housing, and used civic contacts to gain a client for a Section 23 turnkey lease project.

Currently the office has five similar projects underway, plus a city health clinic, and a service center for the power and light company. Along the way, his civic activity led to an appointment to the local planning commission.

Last year Garcia brought in Karl Swope, a Texas A&M classmate ('68), as a partner and they now operate with 3½ professional people and a secretary. Gross fees of Garcia/Swope, Architects, this year will be over $150,000 plus equity in several of the housing projects. The current marketing budget includes $3,500 for the firm's first professionally-printed brochure.

"We've been playing it by ear in many areas in getting the practice started," Garcia says, "but it just didn't seem to make sense not to have a budget for marketing. Basically, we've been spending what we set aside for the effort, and it keeps us moving in good directions."

24

MEASURING
RESULTS

"I don't accept that anyone in business development should go to work without knowing how they are going to be judged."
—Marketer William Hankinson

Waiting to measure marketing results until contracts are signed is far too late in the marketing sequence to assume that the process is going well. The typical lead time from the start of a cold contacting process until leads are found, a selection is made, and work is under contract and ready to start is anywhere from 9 to 18 months. In a firm where several different professionals and/or marketers are pursuing work, it is essential to have yardsticks by which to measure marketing activity well before contracted work is expected in house. Without such a means of measuring marketing progress it is all too easy to wake up at the end of a marketing year and find that little or none of the agreed tasks were performed and the firm is a year behind in filling its marketing pipeline.

There are a number of measurements that can be applied to track ongoing marketing activity. If there is a plan, and those with assigned marketing roles know what they are expected to do, a marketing manager or coordinator can learn in five minutes per week per person whether that activity is taking place. Obviously, the very act of asking once a week what each marketing person has done can in itself be an important generator of activity. But it is equally important for everyone to know what is being measured, because many professionals with part-time marketing roles can get quite discouraged if they think the only standard of measurement is fees actually sold.

Some of the most effective and practical units for measuring progress are:

Marketing Time. As mentioned in the previous chapter, 70–80% of the marketing cost in design firms is personnel time. It follows that if the time is not being expended, the marketing will not get done. One of the most basic controls over marketing performance, therefore, can be through the budgeting and monitoring of time, especially among those key professional who have part-time marketing assignments. If each person in the marketing plan budgets the time they will devote to the responsibility (e.g., half-a-day-a-week to bird-dogging, one day a month to market research, etc.) the marketing manager can then monitor actual time spent to see if these commitments are kept.

List Preparation. Every closer and lead finder in the firm should have a defined list of prospects to contact. This may range from a small list of past clients to be monitored by a technical principal to a large list of strangers prepared by a lead finder trying to open a new market. The marketing manager should know that each person with a marketing role has prepared a list, and should review and approve who is on the lists.

Contact Frequency. The next measure of performance can be the rate at which persons on the list are contacted. It is not of concern that one closer agrees to make one contact per week, while another agrees to make only one a month, while the lead finder is making three a day. Each marketer should be allowed to set their own pace and quota. But each should have a quota, so the manager has a yardstick by which to monitor performance.

Leads Found. The number of leads found from any particular number of contacts is not a valid measure of the person making the contacts. But it can speak to the quality of the list being pursued, and the manager may suggest revising lists if the yield is not productive. Overall, the manager should keep count of total number of live leads known to the firm. As this list grows and shrinks, it is an excellent barometer of the total prospecting effort—and of the prospects for eventual sales.

Size of Leads. Sometimes a lead finder will produce a high volume of live prospects, but they include too many small feasibility studies and other cats and dogs that do not justify the marketing effort. In such cases, it is often practical to set a lead finder's goal in terms of projects above a certain minimum fee, and to manage the list preparation and contacting effort on that basis.

Short Lists from Leads. A measure of the quality of courting that is being performed, and another measure of the quality of leads, is the ratio of times the firm is short-listed in relation to total leads pursued. This is a particularly helpful factor in keeping control of the marketing budget, because the cost of pursuing a lead, including time to court the decision-makers plus submittal of a detailed brochure of qualifications, can easily run into thousands of dollars. If the firm is not making a realistic share of shortlists, the manager may be able to set more selective standards for which leads to pursue seriously.

Batting Average. Mathematically, a firm is breaking even on its marketing effort if it wins jobs in direct proportion to the number of competitors on the short-lists. In other words, if there are five firms on the typical short list, a 20% success rate would be break-even. Keeping score on the firm's yield from competitive situations is a good measure of how competitive the firm is in terms of either its tactical selling strategies, its basic professional capabilities, or both.

The importance of having marketing yardsticks such as these cannot be overemphasized. In the typical design practice marketing is always working toward mid-range and long-range targets. If the firm waits to measure success until those targets are achieved, it is far too late to avoid the penalties of short-fall. Marketing performance should be measurable—and measured—at least quarterly in small firms, and more frequently in larger organizations.

FORECASTING MARKETING ACTIVITY

In addition to the measurements of marketing activity noted above, it is valuable to have some method of forecasting the results of a marketing program far enough in advance so there can be some basis for anticipating the

MARKETING ACTIVITY FORECAST				
Prospect	Anticipated fee (a)	Probability factor (b)	Extension (c)	Decision timetable (d)
XYZ Office Building	$180,000	75%	$135,000	October
GSA Post Office	50,000	20%	10,000	November
ABC Detention Center	500,000	10%	50,000	January
TOTAL	(d)	(d)	$195,000(c)	

Fig. 24-1. Example marketing activity forecast.

firm's workload and the need for new work. Figure 24-1 illustrates a marketing activity forecast which some firms find can produce surprisingly accurate projections of fees from three months to a year ahead.

Recognizing that nothing is ever close to 100% certain in marketing, the forecast relies on the principles of probability judgments for its projections.

The matrix is simply a list of all prospective projects which have a reasonably close timetable for decision, usually 3, 6, or 12 months, depending on the type of work the firm does and its rate of turnover. The forecasting elements (see Fig. 24-1) are:

(a) Estimated fee. Be realistic, and list only the fee from the immediate scope of work. If you are pursuing a $20 million project for which the first phase is a $30,000 feasibility study, that is the only fee to list. After the study is done, and the job is awaiting financing, it can go back on the list with a new timetable and a new probability.

(b) Probability factors. This is always a judgment call, but experienced marketers get quite good at it with a little practice. Of the examples used, the Office Building could be a project where you've been told you have the commission, but instinct tells you there is only a 75% chance the job is real. On the Post Office, a 20% probability would be appropriate if you are one of 5 on the shortlist. In the case of the Prison, a 1-in-10 long shot is worth listing because the goal is to play on the law of averages.

(c) The total of the extended probabilities is the important number. Management can add this number to the backlog of work in progress and get a very good look at the size of the future workload.

(d) As a byproduct, the marketing manager can also compare the average fee and average probability over time to get an idea of what is happening to the size and quality of the leads the firm is pursuing.

The reliability of the activity forecast is, of course, greater as the number of prospects grows. Thus large firms (with long lists) find they can predict billings within 85% to 90% reliability using the matrix approach. Some even have it computerized. The exercise, however, is equally valid for small firms because it can give marketers a disciplined approach to monitor the quality—and the results—of the marketing effort.

SUMMARIZING RESULTS

Periodically during the year it is important to take stock and summarize the marketing activity to date. This exercise produces a valuable look at what has been accomplished to date and, equally important, can reveal the need for adjustments in the Marketing Plan for the period ahead.

How this summary is done is less important than the act of doing it. Fig. 24-2 is a reproduction of an actual marketing memorandum prepared a few years ago by the marketing manager of a 40-person midwest firm of architects, engineers and planners. Fig. 24-3 is a sales analysis matrix patterned after a format developed by a larger, 100-person architecture and engineering firm. The latter format is a good model to both report overall results and provide a valuable analysis of the different sources and types of work being obtained. Any other reporting form that gives the firm's managers a look at marketing results would be equally valuable.

The important point is to have the means of measurement. Marketing that is being measured against a predetermined plan is being made to happen.

MEMORANDUM
TO: Marketing Management Team
FROM: Director of Business Development
SUBJECT: Marketing Results Fiscal Year 1977

New Fees: $
                    59% of new fees were the result of direct sales effort.
                    32% of new fees came from repeat clients.
                    9% of new fees were the result of referrals.
Success Ratio: 31% of the time, were successful when given the oppor-
                    tunity to give a presentation.
                    36% of the time, were successful when a proposal or
                    response was made.
                    14% of new contacts became prospects.
                    19% of new prospects identified became projects.

Direct Sales Effort: 7 new projects; 59% of total new fees.
Repeat Clients: 6 new projects; 32% of total new fees.
Referrals: 4 new projects; 9% of total new fees.

New Contacts Made Fiscal Year 1977:   623
Follow-up Contacts Made Fiscal Year 1977:   44
                                           1,292 total contacts

Prospects Identified Fiscal Year 1977:   84
Prospect Backlog Fiscal Year 1977:   44
                                             128 total prospects

Prospects Identified in Fiscal Year 1977 (Won):   16
Prospects from Fiscal Year 1976 Backlog (Won):   1
                                                  17 prospects won

Presentations: 13
  Won: 4
  Lost: 9
Proposals, responses made: 18
  Won: 10
  Lost: 8

Fig. 24-2.

SALES ANALYSIS				
A. FEES SOLD BY SOURCE			YEAR _____	
	1st Quarter	2nd Quarter	3rd Quarter	4th Quarter
1. Additional Sales on Work in Process				
Project A	$_____		$_____	
Project B		$_____		$_____
Project C				$_____
Etc.				
Sub Total	$_____	$_____	$_____	$_____
2. Repeat Client Sales				
Project D		$_____		$_____
Project E		$_____		
Project F			$_____	$_____
Etc.				
Sub Total		$_____	$_____	$_____
3. New Client Sales				
Project G	$_____			
Project H	$_____			
Project I			$_____	
Project J				$_____
Sub Total	$_____		$_____	$_____
GRAND TOTALS	$_____	$_____	$_____	$_____

B. MARKETING SUCCESS RATE   PERIOD _____

Activity	MARKET A	MARKET B	ALL OTHER	TOTALS
Proposals Submitted	#____	#____	#____	#____
Interviews	#____ (___%)	#____ (___%)	#____ (___%)	#____ (___%)
Awards	#____ (___%)	#____ (___%)	#____ (___%)	#____ (___%)

Fig. 24-3. Sample sales analysis matrix.

SALES ANALYSIS

C. FEES SOLD BY MARKET   PERIOD _____

	Fees Sold	% of Volume
MARKET A		
_____ Projects	$_____	%
MARKET B		
_____ Projects	$_____	%
MARKET C		
_____ Projects	$_____	%
ALL OTHER		
_____ Projects	$_____	%
TOTALS	$_____	100%

D. PROJECTS SOLD BY FEE RANGE

Fee Range	# of Projects	Total Fees
Below $100K	_____	$_____
$100–$250K	_____	$_____
$250–$500K	_____	$_____
$500–$1,000K	_____	$_____
Over $1,000K	_____	$_____
TOTAL	_____	$_____

E. BACKLOG STATUS REPORT   PERIOD _____

PROJECTS	Contracted	Notified of Selection But Not Yet Contracted	On Hold	Cancelled or Fallen Through
Project A	$_____			
Project B	$_____			
Project C	$_____			
Project D		$_____		
Project E	$_____			
Project F	$_____			
Project G	$_____			
Project H			$_____	
Project I				$_____
Project J	$_____			
Etc.				
TOTALS				

Fig. 24-3 (*Continued*)

from: **the Coxe Letter**

CASE STUDY #34: THE ROLE OF INTERVIEWS IN HELPING GET WORK

To shed some light on several topical aspects of marketing architectural services, a questionnaire was distributed among architects attending the marketing seminars at the recent AIA national convention at Atlanta. 106 replies were received: 29 from firms of five people or less; 40 from firms of 6 to 15 people; 23 from firms of 16 to 40 people; and 14 from firms of 40 or more people.

A basic question asked firms to identify how they obtain work (by number of projects / commissions). The tabulation:

Firm size (# of people):	(1–5)	6–15)	(16–39)	(40+)
Repeat Commissions from former clients	43.1%	44.1%	45.2%	45.4%
Direct referral—little or no competition	35.0%	32.6%	19.4%	19.8%
New business from marketing	21.9%	23.3%	35.4%	34.8%
	100%	100%	100%	100%

The results show clearly that in spite of greater competition, the best marketing system is still satisfactory performance which generates repeat and referral work. Size of a firm apparently has nothing to do with the ratio of repeat work. It is interesting that larger firms obtain less of their work from referral: Perhaps it illustrates the value of active bird-dogging in turning up clients, who, had they not been directly solicited, would have picked their architect by referral.

To obtain this new work, respondents to the survey reported the following breakdown of the number of presentations / interviews in which they participate:

Firm size: (# of people):	(1–5)	(6–15)	(16–39)	(40+)
Annual number of informal presentations	17	30	32	45
Annual number of formal interviews	12	11	16	26

This table reveals an apparently great disparity in the yield obtained from interviews / presentations by smaller firms as compared to larger offices. It either indicates that smaller firms are more inclined to compete over their heads (for jobs that would go to a large office in any event); or possibly that large firms are more sophisticated and effective at the interviewing they do.

Another revealing aspect of the survey was the breakdown of total workload and annual interview activity according to major market areas. The results:

Firm size: (# of people):	(1–5)	(6–15)	(16–39)	(40+)
Corporate/Industrial				
% of workload	24%	28%	21%	32%
Informal presentations	1	8	7	11
Formal interviews	2	2	2	7
Institutional				
% of workload	24%	30%	40%	32%
Informal presentations	4	8	10	15
Formal interviews	3	4	8	7
Government				
% of workload	14%	17%	20%	20%
Informal presentations	2	5	9	8
Formal interviews	2	3	4	6
Developer				
% of workload	26%	18%	15%	10%
Informal presentations	6	6	5	4
Formal interviews	3	1	1	2
Other Markets				
% of workload	12%	7%	4%	6%
Informal presentations	4	3	1	7
Formal interviews	2	1	1	4

There is a closer correlation than might be expected between number of interviews and workload. Government work, for example, does not appear to require proportionately more presentations than any other market.

Perhaps the most interesting aspect of the survey is the obvious importance of informal presentations as a marketing tool. While the profession has been preoccupied for years with the mechanics of formal interviews (structured by the client), it is apparent that the informal presentation, where the architect controls the ground rules, can be a very effective way to win commissions.

The survey should encourage everyone about the state of marketing in architectural practice. On one hand it is clear that active, professional business development efforts are producing work in every market for every size practice. At the same time, it is equally clear that there is no better way to retain and obtain clients than to provide outstanding professional service.

V

Where Do We Go From Here?

25

HAVING THE
ANSWERS
IS NOT THE ONLY
ANSWER

"What would happen if every professional were successful in putting the foregoing into practice?"
Asked in the first edition of *Marketing Architectural
and Engineering Services*

The first edition of this book ten years ago spoke to "lifting the veil of mystery" that had surrounded the marketing of professional services for generations.

The next decade saw a virtual marketing revolution in design practice, as engineers, architects, and other professionals learned "how to do it." By the end of the 1970s, however, the body of experience and literature dealing with how to find leads, develop proposals, make presentations, etc. was sufficient for the needs of most firms. At this juncture there was a growing discovery in firm after firm that knowing how to sell professional services did not at all guarantee a successful marketing program. Marketing is always concerned with *what you sell* as well as *how you sell it.* This author's observation is that success in marketing of professional services is weighted by the following factors:

5–10% *Where You Sell It.* The choice of market, both by client / service type, or geographically, is only of major importance in saturated markets. For the most part, there is room in every market for a better professional service.

15–30% *How You Sell It.* Clients are showing an uncanny ability to recognize professionalism, in spite of the paranoia of architects and engineers that no one really understands what they do. Thus, how you sell is important mostly when the competition gets down to a short list of equally qualified firms.

60–80% *What You Sell.* If the professional quality of what you do isn't qualified to put you on the short list, all the selling in the world won't often win the job. Thus, marketing in a design firm involves a great deal more than supervising the job-by-job selling activity.

The question for the future is not: When all design firms know how to sell their services, what will be left for the client to decide who will get the job?

It seems clear from the experience of the past decade that the work is going to those firms that pay more and more attention to what they are selling. The author believes that this is of genuine benefit to both clients and the architects, engineers, and other professionals who serve them. On the one hand design firms are striving to improve and expand the services they offer in order to be competitive. In response clients are learning more and more about the full nature of services design professionals can perform, and are applying those services to even better projects.

Where do we go from here?

Learning marketing techniques has now become a standard consideration of professional practice and will in due course be incorporated into the education and consciousness of future architects and engineers as well. At the same time the vital support role that can be played by marketing staff has become as basic in design firms as specifications writing or business administration. All concerned can now get on with the job of making the architectural and engineering services themselves as responsive as possible to client needs.

In that respect, the answer to the question asked at the beginning of this chapter remains exactly the same as it was at the end of the first edition.

In the face of the enormous projected demand for design services over the remainder of this century, it seems clear that effective marketing programs will not only help individual practitioners prosper, but will also play a vital role in helping the professionals as a whole reach their goals for society.

from: **the Coxe Letter**

CASE STUDY #50. HOW NOT TO TURN DOWN WORK

An award-winning architect in a southern state, whose practice attracts much of its work by reputation and referral, recently demonstrated that it is still professional performance that sells best.

The architect received a telephone call one day from a man who identified himself as part of an architect evaluation team for a public agency that was going to build a $1 million facility in the community. He said he wanted to schedule a visit to the architect's office to interview the firm.

"I was surprised, and suggested that maybe somebody had made a mistake," recalls the architect. "We had not applied for the commission and I thought perhaps he had confused us with another architect with the same name."

The caller confirmed the address and asked, "You are an architect, aren't you?"

"Yes we are," was the reply.

"Well," said the official, "do you mean you aren't interested in an interview?"

"I said we were awfully busy and couldn't consider the present project," reports the architect, who tries to keep his office small (11 people) by design, "but I told him we were interested in future work with the agency and it would be all right if he came by for future reference. He assured me this would be fine."

When the official and a colleague arrived at the office several days later, the architect reports, "I was personally in the midst of something, so I took them around quickly, gave them documents on one of our jobs, and then turned them over to our secretary to get information they needed for their questionnaire. She answered their questions and then they left."

A week later the architect got a phone call from another official of the same agency who said the firm had been selected for the job.

"I said I was sure there was a misunderstanding," the architect recalls, "and asked to talk to the men who had visited the office. Frankly, I thought somebody might be putting me on. But the caller explained that the people who visited us had nothing to do with the final selection. They only made an objective analysis and turned over their score sheets to others."

"We had the impression that to withdraw then would reflect adversely on us," the architect reports, "and we are interested in the client on a long term basis. So we took it."

Appendices

APPENDIX A

BIBLIOGRAPHY

Managing Architectural and Engineering Practice, by Weld Coxe. (1980 John Wiley & Sons. New York)

How to Market Professional Design Services, by Gerre L. Jones. (1973 McGraw Hill. New York)

How to Prepare Professional Design Brochures, by Gerre Jones. (1976 McGraw Hill. New York)

Preparing Design Office Brochures: A Handbook, by David Travers. (1978 Management Books, Santa Monica, CA)

Creative Communications for a Successful Design Practice, by Stephen A. Kliment, AIA. (1977 Whitney, New York)

Architectural and Engineering Salesmanship, by David G. Cooper, AIA. (1978 John Wiley & Sons. New York)

The Design and Building Industry's Publicity Directory, Barbara Welanetz, editor. (1982 A/E Marketing Journal, Box 11316, Newington, CT. 06111)

The Design and Building Industry's Awards Directory, Barbara Welanetz, editor. (1982 A/E Marketing Journal, Box 11316, Newington, CT. 06111)

The Negotiating Game, by Chester L. Karrass. (1970 Thomas Y. Crowell Company, New York)

The Art of Negotiating, by Gerard I. Nierenberg, (1968 Hawthorne Books)

NEWSLETTERS

A/E Marketing Journal (formerly the Coxe Letter) Monthly. Michael R. Hough, Editor, P. O. Box 11316, Newington, CT 06111

Professional Services Management Journal. Monthly. Michael R. Hough, Publisher. P. O. Box 11316, Newington, CT 06111

Professional Marketing Report. Monthly. Gerre Jones, Publisher. P. O. Box 32387, Washington, D.C. 20007

The Presentation Advisor. Bi-monthly. Ernest Burden, Publisher. 20 Waterside Plaza, New York, NY 10010

APPENDIX B

USEFUL REFERENCES AND DIRECTORIES

If you look hard enough, you can find a source or directory on almost any subject imaginable. Following are the titles of a handful of references which can be useful openers to the marketer seeking sources of market data, names to bird-dog or contact for research.

A few hours in a good public library should reassure any doubter about the value of such research. For those who feel the need for instruction in how to use these directories, an excellent, readable text is *Sources of Business Information* by Edwin T. Cowan, Jr. (1964 Revised Edition). U. of California. Another basic tool is the *Guide to American Directories,* B. Klein Publications, P. O. Box 8503, Coral Springs, Fla. 33065.

Professional Reference

The Society for Marketing Professional Services, 1437 Powhatan Street, Alexandria, VA 22314. The professional association of 2,000-plus marketers in design firms; sponsors professional meetings; publishes a newsletter. An essential point of peer contact for anyone in the field.

General References

Encyclopedia of Associations published by Gale Research Company, Book Tower, Detroit, Mich. 48226. Lists National Associations of the United States in business, professional, governmental and related fields, with descriptions of their scope and activities, headquarters, publications, names of executives. Key Word Index provides immediate fix on fields ranging from Airports to Zinc. Excellent tool for locating knowledgeable sources for market research.

Sources of State Information and State Industrial Directories. Chamber of Commerce of the United States, Washington, D.C. Pamphlet containing names and addresses of private and public agencies which publish information about their states. State industrial directories can be extremely helpful in local bird-dogging.

Dun & Bradstreet Million Dollar Directory. Annual. Dun & Bradstreet, New York. Lists all manufacturing firms with a net worth of a million dollars or more. Provides company name, products, plant locations, lists of key offices, etc. A good starter source in the industrial field.

Mac Rae's State Industrial Directories, 87 Terminal Drive, Plainview, NY 11803. State by state volumes listing companies, key executives, and purchasing personnel. Useful for bird-dogs who know what industries they want to cover.

The Municipal Yearbook, International City Management Association, 1140 Connecticut Avenue, N.W., Washington, D.C. 20036. Gives statistical data and names of key public officials in every community over 2500 population in U.S. and Canada.

American Universities and Colleges. American Council on Education, Washington, D.C. Gives data on history, size, curriculum and physical plant of all accredited colleges and universities.

Yearbook of Higher Education. Academic Media, Inc. Los Angeles. Up-to-date list of names of top officials of 2600 colleges and universities.

American Hospital Directory. Annual. American Hospital Association, Chicago. A geographical listing of hospitals and related medical institutions.

Foundation Directory, Russell Sage Foundation, New York. A fascinating compilation of the assets and fields of interest of 6000 foundations. Excellent as a source for seeking direct research grants, or helping clients raise capital funds.

Telephone Directory

Who's Who in America. A. N. Marquis Co., Chicago.

Who's Who in Commerce and Industry. A. N. Marquis Co., Chicago. (Similar specialized directories are published in a number of other fields and geographic territories.)

Poor's Register of Corporation Directors. Standard and Poors Corporation, 345 Hudson St., New York, NY 10014. Alphabetical and organizational cross index of the directors of all major U.S. corporations. Invaluable aid for cross-checking who-knows-whom at the upper strata of the business world.

Ulrich's International Periodicals Directory, R. R. Bowker Company, 1180 Avenue of the Americas, New York, NY 10036. Trade journalists are often the best source of referral to important information. This volume lists the editor or publisher for each of 60,000 periodicals, referenced by subject.

Federal Government

Congressional Directory. Government Printing Office, Washington, D.C. Fairly detailed biographical information on members of Congress and top-rank appointees in all government departments and agencies. (Similar directories are published at most city, county and state government levels.)

U.S. Government Organization Manual. Federal Register Division, U.S. General Archives, Washington, D.C. Can sometimes be helpful in locating the right individual below the top rank in government offices, and is useful for understanding the hierarchy.

A Researcher's Guide to Washington, Washington Researchers, 910 Seventeenth St., N.W., Washington, D.C. 20006. A very helpful guide to how to get information from and through the Federal Government.

APPENDIX C

U.S. Government Standard
Forms 254 and 255—
Architect-Engineer
Questionnaires

| STANDARD FORM (SF) **254** | Architect-Engineer and Related Services Questionnaire | Standard Form 254 General Services Administration, Washington, D. C. 20405 Fed. Proc. Reg. (41 CFR) 1-16. 803 Armed Svc. Proc. Reg. 18-403 |

Purpose:

The policy of the Federal Government, in procuring architectural, engineering, and related professional services, is to encourage firms lawfully engaged in the practice of those professions to submit annually a statement of qualifications and performance data. Standard Form 254, "Architect-Engineer and Related Services Questionnaire" is provided for that purpose. Interested A-E firms (including new, small, and/or minority firms) should complete and file SF 254's with each Federal agency and with appropriate regional or district offices for which the A-E is qualified to perform services. The agency head for each proposed project shall evaluate these qualification resumes, together with any other performance data on file or requested by the agency, in relation to the proposed project. The SF 254 may be used as a basis for selecting firms for discussions, or for screening firms preliminary to inviting submission of additional information.

Definitions:

"Architect-engineer and related services" are those professional services associated with research, development, design and construction, alteration, or repair of real property, as well as incidental services that members of these professions and those in their employ may logically or justifiably perform, including studies, investigations, surveys, evaluations, consultations, planning, programming, conceptual designs, plans and specifications, cost estimates, inspections, shop drawing reviews, sample recommendations, preparation of operating and maintenance manuals, and other related services.

"Parent Company" is that firm, company, corporation, association or conglomerate which is the major stockholder or highest tier owner of the firm completing this questionnaire; i.e. Firm A is owned by Firm B which is, in turn, a subsidiary of Corporation C. The "parent company" of Firm A is Corporation C.

"Principals" are those individuals in a firm who possess legal responsibility for its management. They may be owners, partners, corporate officers, associates, administrators, etc.

"Discipline", as used in this questionnaire, refers to the primary technological capability of individuals in the responding firm. Possession of an academic degree, professional registration, certification, or extensive experience in a particular field of practice normally reflects an individual's primary technical discipline.

"Joint Venture" is a collaborative undertaking by two or more firms or individuals for which the participants are both jointly and individually responsible.

"Consultant", as used in this questionnaire, is a highly specialized individual or firm having significant input and responsibility for certain aspects of a project and possessing unusual or unique capabilities for assuring success of the finished work.

"Prime" refers to that firm which may be coordinating the concerted and

complementary inputs of several firms, individuals or related services to produce a completed study or facility. The "prime" would normally be regarded as having full responsibility and liability for quality of performance by itself as well as by subcontractor professionals under its jurisdiction.

"Branch Office" is a satellite, or subsidiary extension, of a headquarters office of a company, regardless of any differences in name or legal structure of such a branch due to local or state laws. "Branch offices" are normally subject to the management decisions, bookkeeping, and policies of the main office.

Instructions for Filing (Numbers below correspond to numbers contained in form):

1. Type accurate and complete name of submitting firm, its address, and zip code.

1a. Indicate whether form is being submitted in behalf of a parent firm or a branch office. (Branch office submissions should list only personnel in, and experience of, that office.)

2. Provide date the firm was established under the name shown in question 1.

3. Show date upon which all submitted information is current and accurate.

4. Enter type of ownership, or legal structure, of firm (sole proprietor, partnership, corporation, joint venture, etc.)

4a. Check appropriate box indicating if firm is minority-owned. (See 41 CFR 1-1.13 or ASPR 1-332.3(e) for definitions of minority ownership.)

5. Branches or subsidiaries of larger or parent companies, or conglomerates, should insert name and address of highest-tier owner.

5a. If present firm is the successor to, or outgrowth of, one or more predecessor firms, show name(s) of former entity(ies) and the year(s) of their original establishment.

6. List not more than two principals from submitting firm who may be contacted by the agency receiving this form. (Different principals may be listed on forms going to another agency.) Listed principals must be empowered to speak for the firm on policy and contractual matters.

7. Beginning with the submitting office, list name, location, total number of personnel and telephone numbers for all associated or branch offices, (including any headquarters or foreign offices) which provide A-E and related services.

7a. Show total personnel in all offices. (Should be sum of all personnel, all branches.)

8. Show total number of employees, by discipline, in submitting office. (If form is being submitted by main or headquarters office, firm should list total employees, by discipline, in all offices.) While some personnel may be qualified in several disciplines, each person should be counted only once in accord with his or her primary function. Include clerical personnel as "administrative."

254-101

Standard Form 254 July 1975
Prescribed By GSA Fed. Proc. Reg. (41 CFR) 1-16.803

1

STANDARD
FORM (SF)
254
Architect-Engineer
and Related Services
Questionnaire

Standard Form 254
General Services Administration,
Washington, D. C. 20405
Fed. Proc. Reg. (41 CFR) 1-16 . 803
Armed Svc. Proc. Reg. 18-403

Write in any additional disciplines – sociologists, biologists, etc. – and number of people in each, in blank spaces.

9. Using chart (below) insert appropriate index number to indicate range of professional services fees received by submitting firm each calendar year for last five years, most recent year first. Fee summaries should be broken down to reflect the fees received each year for (a) work performed directly for the Federal Government (not including grant and loan projects) or as a sub to other professionals performing work directly for the Federal Government; (b) all other domestic work, U. S. and possessions, including Federally-assisted projects; and (c) all other foreign work.

Ranges of Professional Services Fees

INDEX		INDEX	
1.	Less than $100,000	5.	$1 million to $2 million
2.	$100,000 to $250,000	6.	$2 million to $5 million
3.	$250,000 to $500,000	7.	$5 million to $10 million
4.	$500,000 to $1 million	8.	$10 million or greater

10. Select and enter, in numerical sequence, **not more than thirty** (30) "Experience Profile Code" numbers from the listing (next page) which most accurately reflect submitting firm's demonstrated technical capabilities and project experience. **Carefully review list.** (It is recognized some profile codes may be part of other services or projects contained on list; firms are encouraged to select profile codes which best indicate type and scope of services provided on past projects.) For each code number, show total number of projects and gross fees (in thousands) received for profile projects performed by firm during past five years. If firm has one or more capabilities not included on list, insert same in blank spaces at end of list and show numbers in question 10 on the form. In such cases, the filled-in listing **must** accompany the complete SF 254 when submitted to the Federal agencies.

11. Using the "Experience Profile Code" numbers in the same sequence as entered in item 10, give details of at least one recent (within last five years) representative project for each code number, up to a **maximum** of thirty (30) separate projects, or portions of projects, for which firm was responsible (Project examples may be used more than once to illustrate different services rendered on the same job. Example: a dining hall may be part of an auditorium or educational facility.) Firms which select less than thirty "profile codes" may list two or more project examples (to illustrate specialization) for each code number so long as total of all project examples does not exceed thirty (30). After each code number in question 11, show (a) whether firm was "P," the prime professional, or "C," a consultant, or "JV," part of a joint venture on that particular project (New firms, in existence less than five (5) years may use the symbol "IE" to indicate "Individual Experience" as opposed to firm experience.); (b) provide name and location of the specific project which typifies firm's (or individual's) performance under that code category; (c) give name and address of the owner of that project (if government agency indicate responsible office); (d) show the estimated construction cost (or other applicable

cost) for that portion of the project for which the firm was primarily responsible. (Where no construction was involved, show approximate cost of firm's work); and (e) state year work on that particular project was, or will be, completed.

12. The completed SF 254 should be signed by a principal of the firm, preferably the chief executive officer.

13. Additional data, brochures, photos, etc. should not accompany this form unless specifically requested.

NEW FIRMS (not reorganized or recently-amalgamated firms) are eligible and encouraged to seek work from the Federal Government in connection with performance of projects for which they are qualified. Such firms are encouraged to complete and submit Standard Form 254 to appropriate agencies. Questions on the form dealing with personnel or experience may be answered by citing experience and capabilities of individuals in the firm, based on performance and responsibility while in the employ of others. In so doing, notation of this fact should be made on the form. In question 9, write in "N/A" to indicate "not applicable" for those years prior to firm's organization.

Standard Form 254 July 1975
Prescribed By GSA Fed. Proc. Reg. (41 CFR) 1-16.803

Experience Profile Code Numbers
for use with questions 10 and 11

001 Acoustics; Noise Abatement
002 Aerial Photogrammetry
003 Agricultural Development; Grain Storage; Farm Mechanization
004 Air Pollution Control
005 Airports; Navaids; Airport Lighting; Aircraft Fueling
006 Airports; Terminals & Hangars; Freight Handling
007 Arctic Facilities
008 Auditoriums & Theatres
009 Automation; Controls; Instrumentation
010 Barracks; Dormitories
011 Bridges
012 Cemeteries (Planning & Relocation)
013 Chemical Processing & Storage
014 Churches; Chapels
015 Codes; Standards; Ordinances
016 Cold Storage; Refrigeration; Fast Freeze
017 Commercial Buildings (low rise); Shopping Centers
018 Communications Systems; TV; Microwave
019 Computer Facilities; Computer Service
020 Conservation and Resource Management
021 Construction Management
022 Corrosion Control; Cathodic Protection; Electrolysis
023 Cost Estimating
024 Dams (Concrete; Arch)
025 Dams (Earth; Rock); Dikes; Levees
026 Desalination (Process & Facilities)
027 Dining Halls; Clubs; Restaurants
028 Ecological & Archeological Investigations
029 Educational Facilities; Classrooms
030 Electronics
031 Elevators; Escalators; People-Movers
032 Energy Conservation; New Energy Sources
033 Environmental Impact Studies, Assessments or Statements
034 Fallout Shelters; Blast-Resistant Design
035 Field Houses; Gyms; Stadiums
036 Fire Protection
037 Fisheries; Fish Ladders
038 Forestry & Forest Products
039 Garages; Vehicle Maintenance Facilities; Parking Decks
040 Gas Systems (Propane; Natural; Etc.)

041 Graphic Design
042 Harbors; Jetties; Piers; Ship Terminal Facilities
043 Heating; Ventilating; Air Conditioning
044 Health Systems Planning
045 Highrise; Air-Rights-Type Buildings
046 Highways; Streets; Airfield Paving; Parking Lots
047 Historical Preservation
048 Hospitals & Medical Facilities
049 Hotels; Motels
050 Housing (Residential, Multi-Family; Apartments; Condominiums)
051 Hydraulics & Pneumatics
052 Industrial Buildings; Manufacturing Plants
053 Industrial Processes; Quality Control
054 Industrial Waste Treatment
055 Interior Design; Space Planning
056 Irrigation; Drainage
057 Judicial and Courtroom Facilities
058 Laboratories; Medical Research Facilities
059 Landscape Architecture
060 Libraries; Museums; Galleries
061 Lighting (Interiors; Display; Theatre, Etc.)
062 Lighting (Exteriors; Streets; Memorials; Athletic Fields, Etc.)
063 Materials Handling Systems; Conveyors; Sorters
064 Metallurgy
065 Microclimatology; Tropical Engineering
066 Military Design Standards
067 Mining & Mineralogy
068 Missile Facilities (Silos; Fuels; Transport)
069 Modular Systems Design; Pre-Fabricated Structures or Components
070 Naval Architecture; Off-Shore Platforms
071 Nuclear Facilities; Nuclear Shielding
072 Office Buildings; Industrial Parks
073 Oceanographic Engineering
074 Ordnance; Munitions; Special Weapons
075 Petroleum Exploration; Refining
076 Petroleum and Fuel (Storage and Distribution)
077 Pipelines (Cross-Country—Liquid & Gas)
078 Planning (Community, Regional, Areawide and State)
079 Planning (Site, Installation, and Project)
080 Plumbing & Piping Design
081 Pneumatic Structures; Air-Support Buildings
082 Postal Facilities

083 Power Generation, Transmission, Distribution
084 Prisons & Correctional Facilities
085 Product, Machine & Equipment Design
086 Radar; Sonar; Radio & Radar Telescopes
087 Railroad; Rapid Transit
088 Recreation Facilities (Parks, Marinas, Etc.)
089 Rehabilitation (Buildings; Structures; Facilities)
090 Resource Recovery; Recycling
091 Radio Frequency Systems & Shieldings
092 Rivers; Canals; Waterways; Flood Control
093 Safety Engineering; Accident Studies; OSHA Studies
094 Security Systems; Intruder & Smoke Detection
095 Seismic Designs & Studies
096 Sewage Collection, Treatment and Disposal
097 Soils & Geologic Studies; Foundations
098 Solar Energy Utilization
099 Solid Wastes; Incineration; Land Fill
100 Special Environments; Clean Rooms, Etc.
101 Structural Design; Special Structures
102 Surveying; Platting; Mapping; Flood Plain Studies
103 Swimming Pools
104 Storm Water Handling & Facilities
105 Telephone Systems (Rural; Mobile; Intercom, Etc.)
106 Testing & Inspection Services
107 Traffic & Transportation Engineering
108 Towers (Self-Supporting & Guyed Systems)
109 Tunnels & Subways
110 Urban Renewal; Community Development
111 Utilities (Gas & Steam)
112 Value Analysis; Life-Cycle Costing
113 Warehouses & Depots
114 Water Resources; Hydrology; Ground Water
115 Water Supply, Treatment and Distribution
116 Wind Tunnels; Research/Testing Facilities Design
117 Zoning; Land Use Studies
201 _____
202 _____
203 _____
204 _____
205 _____

Standard Form 254 July 1975
Prescribed By GSA Fed. Proc. Reg. (41 CFR) 1-16.803

STANDARD FORM (SF)
254

Architect-Engineer
and Related Services
Questionnaire

1. Firm Name / Business Address:

2. Year Present Firm Established:

3. Date Prepared:

1a. Submittal is for ☐ Parent Company   ☐ Branch Office

4. Type of Ownership:

4a. Minority Owned ☐ yes   ☐ no

5. Name of Parent Company, if any:

5a. Former Firm Name(s), if any, and Year(s) Established:

6. Names of not more than Two Principals to Contact: Title / Telephone

1)

2)

7. Present Offices: City / State / Telephone / No. Personnel Each Office

7a. Total Personnel _____

8. Personnel by Discipline:

_____ Administrative
_____ Architects
_____ Chemical Engineers
_____ Civil Engineers
_____ Construction Inspectors
_____ Draftsmen
_____ Ecologists
_____ Economists
_____ Electrical Engineers
_____ Estimators
_____ Geologists
_____ Hydrologists
_____ Interior Designers
_____ Landscape Architects
_____ Mechanical Engineers
_____ Mining Engineers
_____ Oceanographers
_____ Planners: Urban/Regional
_____ Sanitary Engineers
_____ Soils Engineers
_____ Specification Writers
_____ Structural Engineers
_____ Surveyors
_____ Transportation Engineers

9. Summary of Professional Services Fees Received: (insert index number)

Last 5 Years (most recent year first)

19 ____   19 ____   19 ____   19 ____   19 ____

Direct Federal contract work, including overseas

All other domestic work

All other foreign work*

*Firms interested in foreign work, but without such experience, check here: ☐.

Ranges of Professional Services Fees

INDEX

1. Less than $100,000
2. $100,000 to $250,000
3. $250,000 to $500,000
4. $500,000 to $1 million
5. $1 million to $2 million
6. $2 million to $5 million
7. $5 million to $10 million
8. $10 million or greater

Standard Form 254 July 1975
Prescribed By GSA Fed. Proc. Reg. (41 CFR) 1-16.803

4

10. Profile of Firm's Project Experience, Last 5 Years

Profile Code	Number of Projects	Total Gross Fees (in thousands)	Profile Code	Number of Projects	Total Gross Fees (in thousands)	Profile Code	Number of Projects	Total Gross Fees (in thousands)
1)			11)			21)		
2)			12)			22)		
3)			13)			23)		
4)			14)			24)		
5)			15)			25)		
6)			16)			26)		
7)			17)			27)		
8)			18)			28)		
9)			19)			29)		
10)			20)			30)		

11. Project Examples, Last 5 Years

Profile Code	"P", "C", "JV", or "IE"	Project Name and Location	Owner Name and Address	Cost of Work (in thousands)	Completion Date (Actual or Estimated)
1					
2					
3					
4					
5					
6					
7					

Standard Form 254 July 1975
Prescribed By GSA Fed. Proc. Reg. (41 CFR) 1-16.803

5

Standard Form 254 July 1975
—scribed By GSA Fed. Proc. Reg. (41 CFR) 1-16.803

8

9

10

11

12

13

14

15

16

17

18

19

6

20							
21							
22							
23							
24							
25							
26							
27							
28							
29							
30							

12. The foregoing is a statement of facts

Signature: _____ Typed Name and Title: _____ Date: _____

Standard Form 254 July 1975
Prescribed By GSA Fed. Proc. Reg. (41 CFR) 1-16.803

7

STANDARD FORM (SF)

255

Architect-Engineer and Related Services Questionnaire for Specific Project

Standard Form 255
General Services Administration,
Washington, D. C. 20405
Fed. Proc. Reg. (41 CFR) 1-16 . 803
Armed Svc. Proc. Reg. 18-403

Purpose:

This form is a supplement to the "Architect-Engineer and Related Services Questionnaire" (SF 254). Its purpose is to provide additional information regarding the qualifications of interested firms to undertake a specific Federal A-E project. Firms, or branch offices of firms, submitting this form should enclose (or already have on file with the appropriate office of the agency) a current (within the past year) and accurate copy of the SF 254 at that office.

The procurement official responsible for each proposed project may request submission of the SF 255 "Architect-Engineer and Related Services Questionnaire for Specific Project" in accord with applicable civilian and military procurement regulations and shall evaluate such submissions, as well as related information contained on the Standard Form 254, and any other performance data on file with the agency, and shall select firms for subsequent discussions leading to contract award in conformance with Public Law 92-582. This form should only be filed by an architect-engineer or related services firm when requested to do so by the agency or by a public announcement. Responses should be as complete and accurate as possible, contain data relative to the specific project for which you wish to be considered, and should be provided, by the required due date, to the office specified in the request or public announcement.

This form will be used only for the specified project. Do not refer to this submittal in response to other requests or public announcements.

Definitions:

"**Architect-engineer and related services**" are those professional services associated with research, development, design and construction, alteration, or repair of real property, as well as incidental services that members of these professions and those in their employ may logically or justifiably perform, including studies, investigations, surveys, evaluations, consultations, planning, programming, conceptual designs, plans and specifications, cost estimates, inspections, shop drawing reviews, sample recommendations, preparation of operating and maintenance manuals, and other related services.

"**Principals**" are those individuals in a firm who possess legal responsibility for its management. They may be owners, partners, corporate officers, associates, administrators, etc.

"**Discipline**", as used in this questionnaire, refers to the primary technological capability of individuals in the responding firm. Possession of an academic degree, professional registration, certification, or extensive experience in a particular field of practice normally reflects an individual's primary technical discipline.

"**Joint Venture**", is a collaborative undertaking of two or more firms or individuals for which the participants are both jointly and individually responsible.

"**Key Persons, Specialists, and Individual Consultants**", as used in this questionnaire, refer to individuals who will have **major** project responsibility or will provide **unusual or unique** capabilities for the project under consideration.

Instructions for Filing (Numbers below correspond to numbers contained in form):

1. Give name and location of the project for which this form is being submitted.

2. Provide appropriate data from the *Commerce Business Daily* (CBD) identifying the particular project for which this form is being filed.

2a. Give the date of the *Commerce Business Daily* in which the project announcement appeared, or indicate "not applicable" (N/A) if the source of the announcement is other than the CBD.

2b. Indicate Agency identification or contract number as provided in the CBD announcement.

3. Show name of the individual or firm (or joint venture) which is submitting this form for the project.

3a. List the name, title, and telephone number of that principal who will serve as the point of contact. Such an individual must be empowered to speak for the firm on policy and contractual matters and should be familiar with the programs and procedures of the agency to which this form is directed.

3b. Give the address of the specific office which will have responsibility for performing the announced work.

4. Insert the number of personnel by discipline presently employed (on date of this form) at work location. While some personnel may be qualified in several disciplines, each person should be counted only once in accord with his or her primary function. Include clerical personnel as "administrative." Write in any additional disciplines — sociologists, biologists, etc. — and number of people in each, in blank spaces.

5. Answer only if this form is being submitted by a joint venture of two or more collaborating firms. Show the names and addresses of all individuals or organizations expected to be included as part of the joint venture and describe their particular areas of anticipated responsibility. (i.e. technical disciplines, administration, financial, sociological, environmental, etc.)

5a. Indicate, by checking the appropriate box, whether this particular joint venture has successfully worked together on other projects.

Each firm participating in the joint venture should have a Standard Form 254 on file with the contracting office receiving this form. Firms which do not have such forms on file should provide same immediately along with a notation

255-101

Standard Form 255 July 1975
Prescribed By GSA Fed. Proc. Reg. (41 CFR) 1-16.803

1

STANDARD FORM (SF)

255

Architect-Engineer and Related Services Questionnaire for Specific Project

Standard Form 255
General Services Administration,
Washington, D. C. 20405
Fed. Proc. Reg (41 CFR) 1-16 . 803
Armed Svc. Proc. Reg. 18-403

regarding their association with this joint venture submittal.

6. If respondent is not a joint venture, but intends to use outside (as opposed to in-house or permanently and formally affiliated) consultants or associates, he should provide names and addresses of all such individuals or firms, as well as their particular areas of technical/professional expertise, as it relates to this project. Existence of previous working relationships should be noted. If more than eight outside consultants or associates are anticipated, attach an additional sheet containing requested information.

7. Regardless of whether respondent is a joint venture or an independent firm, provide brief resumes of key personnel expected to participate on this project. Care should be taken to limit resumes to only those personnel and specialists who will have major project responsibilities. Each resume must include: (a) name of each key person and specialist and his or her title, (b) the project assignment or role which that person will be expected to fulfill in connection with this project, (c) the name of the firm or organization, if any, with whom that individual is presently associated, (d) years of relevant experience with present firm and other firms, (e) the highest academic degree achieved and the discipline covered (if more than one highest degree, such as two Ph.D.'s, list both), the year received and the particular technical/professional discipline which that individual will bring to the project, (f) if registered as an architect, engineer, surveyor, etc., show only the field of registration and the year that such registration was first acquired. If registered in several states, do not list states, and (g) a synopsis of experience, training, or other qualities which reflect individual's potential contribution to this project. Include such data as: familiarity with Government or agency procedures, similar type of work performed in the past, management abilities, familiarity with the geographic area, relevant foreign language capabilities, etc. Please limit synopsis of experience to directly relevant information.

8. List up to ten projects which demonstrate the firm's or joint venture's competence to perform work similar to that likely to be required on this project. The more recent such projects, the better. Prime consideration will be given to projects which illustrate respondent's capability for performing work similar to that being sought. Required information must include: (a) name and location of project, (b) brief description of type and extent of services provided for each project (submissions by joint ventures should indicate which member of the joint venture was the prime on that particular project and what role it played), (c) (i) name and address of the owner of that project (if Government agency, indicate responsible office), (d) completion date (actual or estimated), (e) total construction cost of completed project, (or where no construction was involved, the approximate cost of your work) and that portion of the cost of the project for which the named firm was/is responsible.

9. List only those projects which the A-E firm or joint venture, or members of the joint venture, are currently performing under direct contract with an agency or department of the Federal Government. Exclude any grant or loan projects being financed by the Federal Government but being performed under contract to other non Federal governmental entities. Information provided under each heading is similar to that requested in the preceding item 8, except for (d) "Percent Complete" Indicate in this item the percentage of A-E work completed upon filing this form.

10. Through narrative discussion, show reason why the firm or joint venture submitting this questionnaire believes it is especially qualified to undertake the project. Information provided should include, but not be limited to, such data as: specialized equipment available for this work, any awards or recognition received by a firm or individuals for similar work, required security clearances, special approaches or concepts developed by the firm relevant to this project, etc. Respondents may say anything they wish in support of their qualifications. When appropriate, respondents may supplement this proposal with graphic material and photographs which best demonstrate design capabilities of the team proposed for this project.

11. Completed forms should be signed by the chief executive officer of the joint venture (thereby attesting to the concurrence and commitment of all members of the joint venture), or by the architect-engineer principal responsible for the conduct of the work in the event it is awarded to the organization submitting this form. Joint ventures selected for subsequent discussions regarding this project must make available a statement of participation signed by a principal of each member of the joint venture. ALL INFORMATION CONTAINED IN THE FORM SHOULD BE CURRENT AND FACTUAL.

Standard Form 255 July 1975
Prescribed By GSA Fed. Proc. Reg (41 CFR) 1-16 803

2

OMB Approval No. 29-R0235

| STANDARD FORM (SF) **255** Architect-Engineer Related Services for Specific Project | 1. Project Name/Location for which Firm is Filing: | 2a. *Commerce Business Daily* Announcement Date, if any: | 2b. Agency Identification Number, if any: |

3. Firm (or Joint-Venture) Name & Address

3a. Name, Title & Telephone Number of Principal to Contact

3b. Address of office to perform work, if different from Item 3

4. Personnel by Discipline:

____ Administrative
____ Architects
____ Chemical Engineers
____ Civil Engineers
____ Construction Inspectors
____ Draftsmen
____ Ecologists
____ Economists
____ Electrical Engineers
____ Estimators
____ Geologists
____ Hydrologists
____ Interior Designers
____ Landscape Architects
____ Mechanical Engineers
____ Mining Engineers
____ Oceanographers
____ Planners: Urban/Regional
____ Sanitary Engineers
____ Soils Engineers
____ Specification Writers
____ Structural Engineers
____ Surveyors
____ Transportation Engineers
____ Total Personnel

5. If submittal is by Joint-Venture list participating firms and outline specific areas of responsibility (including administrative, technical and financial) for each firm: (Attach SF 254 for each if not on file with Procuring Office.)

5a. Has this Joint-Venture previously worked together? ☐ yes ☐ no

Standard Form 255 July 1975
Prescribed By GSA Fed. Proc. Reg. (41 CFR) 1-16.803

6. Outside Key Consultants/Associates Anticipated for this Project (Attach SF 254 for Consultants/Associates Listed, if not already on file with the Procuring Office)

Name & Address	Specialty	Worked with Prime before (Yes or No)
1)		
2)		
3)		
4)		
5)		
6)		
7)		
8)		

Standard Form 255 July 1975
Prescribed By GSA Fed. Proc. Reg. (41 CFR) 1-16.803

7. Brief Resume of **Key** Persons, Specialists, and Individual Consultants Anticipated for this Project

a. Name & Title:

b. Project Assignment:

c. Name of Firm with which associated:

d. Years experience: With This Firm _____ With Other Firms _____

e. Education: Degree(s)/Year/Specialization

f. Active Registration: Year First Registered/Discipline

g. Other Experience and Qualifications relevant to the proposed project:

a. Name & Title:

b. Project Assignment:

c. Name of Firm with which associated:

d. Years experience: With This Firm _____ With Other Firms _____

e. Education: Degree(s)/Year/Specialization

f. Active Registration: Year First Registered/Discipline

g. Other Experience and Qualifications relevant to the proposed project:

Standard Form 255 July 1975
Prescribed By GSA Fed. Proc. Reg. (41 CFR) 1-16.803

8. Work by Firm or Joint Venture Members which **Best Illustrates Current Qualifications Relevant to this Project** (List not more than 10 Projects)

a. Project Name & Location	B. Nature of Firm's Responsibility	c. Owner's Name & Address	d. Completion Date (actual or estimated)	e. Estimated Cost (in thousands)	
				Entire Project	Work for which Firm was/is responsible
(1)					
(2)					
(3)					
(4)					
(5)					
(6)					
(7)					
(8)					
(9)					
(10)					

Standard Form 255 July 1975
Prescribed By GSA Fed. Proc. Reg. (41 CFR) 1-16.803

9. All work by firms or Joint Venture members currently being performed **directly for Federal agencies**

a. Project Name & Location	b. Nature of Firm's Responsibility	c. Agency (Responsible Office) Name & Address	d. Percent complete	e. Estimated Cost (In Thousands)	
				Entire Project	Work for which firm is responsible

Standard Form 255 July 1975
Prescribed By GSA Fed. Proc. Reg. (41 CFR) 1-16.803

10

10. Use this space to provide any additional information or description of resources supporting your firm's qualifications for the proposed project

11. The foregoing is a statement of facts.

Signature: _____ Typed Name and Title: _____

Standard Form 255 July 1975
Prescribed By GSA Fed. Proc. Reg. (41 CFR) 1-16.803

APPENDIX D

APPENDIX D

The following document was initially developed as a proposal of services by Rahenkamp Sacks Wells and Associates, Land Planners, of Philadelphia. It is cited as an excellent example of a selling proposal.

STANDARD SERVICES*

I. PLANNING SERVICES

Planning Services are intended to carry the project from early feasibility and fact finding through the necessary planning steps to obtain zoning approvals and financing commitments. The fee basis for these services recognizes the speculative nature of the investment at this time, and the desire to keep commitments to a minimum, but permits adequate service to assure that conclusions are realistic for eventual design development and construction.

REGIONAL LAND USE, MARKET SURVEYS AND SITE SELECTION

Collection and analysis of existing natural (water, slope, soils) physical (transportation, land use, utilities) and social conditions (markets, preferences and characteristics) on a regional basis to develop alternate areas of specific interest. Results presented in written reports, drawings and conferences with client for final site selection.

FEE: Negotiated Sum

SITE EVALUATION AND FEASIBILITY

Analysis of natural, physical and social determinants of the selected site to provide comparative recommended development programs and preliminary yield projections.
Services include:
- a. Sub-regional analysis plans and report
- b. Site analysis plans and report
- c. Permissible land use plan(s)

---

*Copyright 1963 Rahenkamp Sachs Wells & Assoc., Inc.

d. Tentative site layout(s)

e. Preliminary informal discussions with local planning boards. (sketch plat)

Produces a documented report and recommendations.

FEE: $60.–$120.00 per acre (minimum $20,000)

THE MASTER PLAN

Proceeding from an approved feasibility, development of a specific Master Plan showing:

a. Layout on site of dwelling units, commercial area, industrial area, open space, roads, parking and utilities to clearly define the general system, technical and economic feasibility.

b. Typical units in outline form to establish coverage, density, mix and general bulk character.

Produces basic documents adequate to submit for public and financial approvals. (Preliminary Plat)

FEE: $40.–$80. per acre (minimum $15,000.)

PUBLIC APPROVALS

Submission of Master Plan for official review and approval by local authorities. Services include all necessary conference and public presentations, but do not include special models, drawings, slides and other extra presentation materials (See Group III—Supplemental Services).

FEE: Negotiated Sum

FINANCING SUBMITTAL

Presentation with or in behalf of the client of the approved Master Plan to financial institutions to establish availability and extent of financing, followed by selection of the first section for development and establishment of reasonably reliable cost estimates to enable the client to negotiate a letter of intention with the financial institution.

FEE: Negotiated Sum

II. DESIGN AND CONSTRUCTION SERVICES: Site

The fee for this group of services is calculated on a unit basis for those units which will be initially constructed. The same unit price will apply to subsequent sections as developed.

PRELIMINARY DESIGN

Preparation of required technical drawings for final plat approval of the first section, and development of preliminary designs for all site work to establish materials and methods in sufficient detail for quantity survey and cost estimate. Conclusion of preliminary design is a set of documents submitted for specific approval by the client as the basis for construction drawings.

FEE: $40.–$60. per unit.

ARCHITECTURAL COORDINATION

Services include weekly conferences with the client and architect to assure coordination of architectural design with the master plan intention, within the economic framework approved by the client and financial institution.

FEE: $10.–$15. per unit.

CONSTRUCTION DRAWINGS

Preparation of technical documents for the site development adequate to obtain building permits and to enable contractors to perform the site work. The following documents, as required, include:

Final grading and surface drainage plan;
Final dimensioned layout plan for buildings;
Final utilities alignment and placement, under ground and above ground;
Final plans, profiles, sections and qualifications for vertical and horizontal road alignments, bridges, etc;
Final landscape and planting plan;
Final construction and planting details;
Specifications.

FEE: $60.–$80. per unit.

INSPECTION

Visits to the site to approve or disapprove work as consistent with drawings and generally check scheduling and procedures. Services include approval of shop drawings, check lists, change orders, progress payments, and material submittals as requested in specifications. Services do not include superintendents, bidding by sub-contractors or day to day coordination of operations. The basic inspection fee allows a maximum of two (2) man hours per unit. When these hours have been used, or one year from the initiation of a particular parcel, lot or section, whichever comes first, additional inspection services thereafter shall be charged on an hourly basis.

FEE: $40.–$45. per unit.

III. SUPPLEMENTARY SERVICES

The common desire for a successful project which we share with the client often indicates the need for supplementary services which are available at the discretion of the client. Rahenkamp Sachs Wells and Associates have successfully performed these services on prior projects and can illustrate their value. An equitable fee basis for these services is a man-hour rate within maximum limits established by prior agreement for each category.

ARCHITECTURAL COORDINATION

Review and comment services beyond the preliminary stage, extending through working drawings and inspection. Not all decisions can be anticipated in the drawings and specifications, therefore, coordination of field architectural decisions is often critical to produce a unified effect between inside and outside spaces.

GRAPHIC DESIGN

Sketches and construction layouts for signs, fences, symbols, brochures, advertising and promotion in order to produce a coordinated design effect.

PRESENTATION MATERIALS AND PROMOTION

The production of brochures, models, sketches, promotional floor plans, photography and other presentation materials for public appearances, marketing and exhibits. Promotion includes presentation of planning and design concepts through interviews, tours, news releases and public appearances, which seek to make the project newsworthy.

INTERIOR DESIGN

Furniture layouts, specification of materials, lighting and color finishes for model units and public spaces including all necessary construction drawings, instructions, sketches, coordination and inspection.

APPENDIX E

JOB AND PERSON DESCRIPTION FOR A MARKETING DIRECTOR/
MANAGER

POSITION   Director/Manager of Marketing

REPORTS TO   (Board/Partnership/or General Manager)

EXTENT OF JOB   Under the general direction of management, plans
                        and coordinates all marketing activities of the firm.
                        Responsibilities involve developing and executing
                        a marketing program which consists of
                        establishing long-range marketing goals and near-
                        term objectives; selecting target markets; and, in
                        close coordination with the principals, screening
                        leads, contacting and cultivating prospects, and
                        obtaining contracts for projects.

SPECIFIC DUTIES

Planning   • Has major responsibility for developing and
                           executing a Marketing Plan which outlines all
                           major aspects of a planned marketing effort
                           including appropriate objectives, tasks,
                           assignments of responsibility to individuals
                           and budgets. The Marketing Plan is to be
                           developed in coordination with the principles
                           approved by management. The remaining
                           specific duties are to be performed within the
                           framework of the marketing plan.

Market Direction   • Determine what specific markets the firm will
                           cultivate. The determination will be made
                           after careful appraisal of the likely market
                           opportunities for the firm.

Targeting   • Establish, for each market where active new
                           business effort is to be applied, specific
                           performance goals for the subsequent twelve
                           months. These goals should include:

(a) The number of new organizations to be contacted during the year.

(b) Where applicable, the number of follow-up calls to be made on former clients or prior contacts.

(c) The number of anticipated interviews with prospects who are actively considering selection of professional services.

(d) The desired volume of commissions and total gross fees to be realized from the sales effort in this market.

Lead Developing

- Operating directly or through designated staff personnel, develop lists of individuals and/or projects (leads/prospects) to be contacted. Organize and screen leads/prospects to produce an efficient procedure for scheduling contacts. Ensure that appropriate individuals are assigned, scheduled, and equipped to make the desired business development contacts.

Strategy Development

- Develop strategies for cultivating prospects and obtaining work at the target market, individual prospect, and project levels. Play a coordinating role in preparing content for interviews and presentations, including determining the level of participation by other staff personnel. Take primary responsibility for coordinating proposals, pricing, and response strategy.

Client Maintenance

- Keep abreast of the progress of projects and see that the firm maintains a continuing relationship with clients during the course of and after the completion of assignments—including project oriented public relations activities.

Sales Tools

- Take primary responsibility for determining what materials and information are needed, coordinating the production of such materials and information. This will include proposals, graphics, brochures, presentation materials, the U.S. Gov't #254, business development oriented correspondence, etc.

Merchandising and Public
Relations

● Develop a plan for appropriate advertising and / or public relations activities to support the firm's marketing program, and see that appropriate staff or consultants are retained to carry it out.

Monitoring

● Maintain (directly or through designated staff personnel) a record and control system to alert appropriate individuals when contacts are to be made; and to serve as a log to record all marketing activities.

Maintain (directly or through designated staff personnel) a control system to measure the performance of the marketing effort; and an activity review system for management.

Marketing Staff

● With the approval of management, hire or obtain the services of others to help carry out the marketing functions.

General

● Keep abreast of issues and current trends in the marketplace and the profession likely to affect the firm and its marketing activities.

PERSONAL QUALITIES

Ability to plan and organize.

Communicates well with others both in face-to-face and in group meetings.

Flexible, readily adaptable to changing conditions and demands.

Self-confident and not easily discouraged.

Able to relate to client and client's concerns and interests. Interested in public and community activities.

Creative, problem-solving approach to work and a facility to generate new ideas and build on the ideas of others.

APPENDIX F

JOB AND PERSON DESCRIPTION FOR LEAD FINDER

POSITION

Lead Finder

REPORTS TO

Director of Marketing

EXTENT OF JOB

Coordinates and monitors on-going contacts and communications between staff and new business leads and prospects. Searches out, identifies and establishes initial contact with leads in new business situations and markets, and assists in client relations with current and past clients. Establishes rapport, monitors and maintains communications, and elicits such information from prospects to allow development of strategy for a successful marketing team approach.

SPECIFIC DUTIES

List Building

- Develop lead lists from various sources suitable to firm markets and potential volume.

Contacting

- Establish contact with leads and elicit such information as to the type of project, dollar volume, schedule, method of financing, decision-making process, decision-maker(s), and other pertinent information. Also coordinate and monitor lead development activities of the staff that may overlap.

Strategy Development

- Based on indicated information, develop strategy that is most promising to present the firm and win project. In concert with Director of Marketing, develop most appropriate marketing team for presentations and formal interviews. Responsible for assembling group and meshing their schedules with those of the project.

Interface

● In addition to client contact for new business development, assist in the researching and preparation of sales presentations. Make effective handoff to members of the marketing team.

Communications

● Internal—Keep Director of Marketing appraised through regularly scheduled reports of actions, schedules and potential prospects.

External—Responsible for coordination of submittals to prospects of all information necessary to reinforce the firm's position, i.e., brochures, 254–255 Forms, articles, leaving pieces, strong "selling" letters, etc.

Market Research

● Keep knowledgeable of all aspects of the profession and current trends in assigned markets. Make special marketing surveys, as required.

PERSONAL QUALITIES

● Confident, poised, ambitious. Not easily discouraged by rejection.

Able to collect, analyze, communicate and act on client data.

Energetic, a doer. Responsive to the firm and able to convey that enthusiasm to potential clients. Warm, open personality. Not dogmatic or inflexible. Readily adaptable to changing condition or demands.

Creative problem-solver; likely to try several approaches when initial solution fails.

APPENDIX G

JOB AND PERSON DESCRIPTION FOR A MARKETING COORDINATOR

POSITION	Marketing Coordinator
EXTENT OF JOB	Coordinate marketing, market research and public relations activities of the firm, under the direction of the (Marketing Director or General Manager) and working closely with all principals. Responsibilities include support of the principals of the firm in their individual lead finding and selling efforts; monitoring the progress of all marketing efforts; organizing the development of all administrative systems and selling tool resources.

SPECIFIC DUTIES

Market Research
- Assist in identifying contacts for sources of market research information and preparing questions to be used in research interviews; compile data gathered during research interviews; help interpret results of research and apply them to lead finding and selling efforts.

Contact Lists
- Assist in developing principals' individual lists of contracts in target markets; transfer raw data into a format for the organization and maintenance of lists; administer annual review of lists to add and delete names and correct data.

Lead Finding
- Assist principals in scheduling contacts and preparing for sales calls; record data and follow-up recommendations, prepare follow-up materials as required, maintain follow-up schedule and assist in the preparation of correspondence.

Response	● As leads are identified, assist principals in planning, coordinating and conducting strategy research efforts; assist in preparing materials for credentials submittals, presentations and proposals.
Selling Operations	● Monitor effectiveness of selling effort and secure feedback from prospects regardless of outcomes. Maintain all marketing paperwork systems including call reports; mailing lists; screening reports; and prospect correspondence files.
Sales Tools	● Organize and maintain photographic, slide and proposal files and data on the credentials of the firm; assist in the development of brochure materials.
Public Relations	● Assist in the development and coordination of special events and feature and routine publicity (in concert with the firm's public relations counsel, if appropriate). Identify opportunities for placement of articles and publicity in professional journals, magazines, and newspapers. Possibly, develop and prepare a periodic firm newsletter.
General	● Remain in close contact with principals on business deyelopment, giving input and guidance as necessary. Perform other marketing and public relations duties as requested by the principals.
PERSONAL QUALITIES	Well-organized and self-disciplined.
	Persistent and assertive but not abrasive.
	Excellent writing abilities.
	Ability to communicate well with others.
	Graphic design sensitivity and taste.
	Public relations orientation.

INDEX

INDEX TO COXE LETTER CASE STUDIES